About Island Press

Since 1984, the nonprofit Island Press has been stimulating, shaping, and communicating the ideas that are essential for solving environmental problems worldwide. With more than 800 titles in print and some 40 new releases each year, we are the nation's leading publisher on environmental issues. We identify innovative thinkers and emerging trends in the environmental field. We work with world-renowned experts and authors to develop cross-disciplinary solutions to environmental challenges.

Island Press designs and implements coordinated book publication campaigns in order to communicate our critical messages in print, in person, and online using the latest technologies, programs, and the media. Our goal: to reach targeted audiences—scientists, policymakers, environmental advocates, the media, and concerned citizens—who can and will take action to protect the plants and animals that enrich our world, the ecosystems we need to survive, the water we drink, and the air we breathe.

Island Press gratefully acknowledges the support of its work by the Agua Fund, Inc., The Margaret A. Cargill Foundation, Betsy and Jesse Fink Foundation, The William and Flora Hewlett Foundation, The Kresge Foundation, The Forrest and Frances Lattner Foundation, The Andrew W. Mellon Foundation, The Curtis and Edith Munson Foundation, The Overbrook Foundation, The David and Lucile Packard Foundation, The Summit Foundation, Trust for Architectural Easements, The Winslow Foundation, and other generous donors.

The opinions expressed in this book are those of the author(s) and do not necessarily reflect the views of our donors.

EVOLUTION IN A TOXIC WORLD

Evolution in a
Toxic World

HOW LIFE RESPONDS TO CHEMICAL THREATS

Emily Monosson

 **ISLAND**PRESS

Washington | Covelo | London

Library of Congress Cataloging-in-Publication Data

Monosson, Emily.
Evolution in a toxic world : how life responds to chemical threats / Emily Monosson.
p. cm.
Includes bibliographical references and index.
ISBN 978-1-59726-976-6 (cloth : alk. paper) — ISBN 1-59726-976-X (cloth : alk. paper) — ISBN 978-1-59726-977-3 (pbk. : alk. paper) — ISBN 1-59726-977-8 (pbk. : alk. paper) 1. Environmental toxicology. 2. Adaptation (Physiology) 3. Ecophysiology. 4. Evolution (Biology) I. Title.
RA1226.M66 2012
613'.1—dc23
2012003365

Printed on recycled, acid-free paper

Manufactured in the United States of America
10 9 8 7 6 5 4 3 2 1

CONTENTS

Preface   ix

Acknowledgments   xi

Chapter 1   An Introduction   1

PART 1 ELEMENT   13
Chapter 2   Shining a Light on Earth's Oldest Toxic Threat?   15
Chapter 3   When Life Gives You Oxygen, Respire   33
Chapter 4   Metal Planet   49

PART 2 PLANT AND ANIMAL   65
Chapter 5   It Takes Two (or More) for the Cancer Tango   67
Chapter 6   Chemical Warfare   83
Chapter 7   Sensing Chemicals   101
Chapter 8   Coordinated Defense   117

PART 3 HUMAN   131
Chapter 9   Toxic Evolution   133
Chapter 10 Toxic Overload?   149

Appendix: Five Recent Additions to the Chemical Handbook of Life   161
Notes   173
Selected Bibliography   211
Index   217

PREFACE

Before I embarked on this journey through time, the word "evolution" called to mind images of finch beaks, squid eyes, and that pervasive lineup of an ape morphing into a human slumped over a computer—an example of a relatively "modern" evolutionary change. I never considered the very long (billions of years) evolutionary history of the systems that I had studied for decades. The proteins and enzymes evolved partially in response to the plethora of chemicals that threaten to upset the balance of life. But as an environmental toxicologist focused on the effects of chemicals today, I never saw it that way. From my contemporary pedestal, I could only see from the top down. I focused solely on the adverse effects of chemical contaminants and, more recently, on the ways chemicals used in industrial and consumer products affect both humans and wildlife. I rarely if ever stopped to ask how we got here, even though for years I have been teaching environmental studies students that they must understand history, not just to understand the present, but to change the future. Now I am taking my own advice. We are faced with a barrage of chemicals both familiar and unfamiliar to life. Truly understanding the effects of these chemicals, and changing the way we create, use, and evaluate them, requires deeper study of life's history of chemical defense.

Acknowledging my own limitations, and feeling much like a graduate student without an adviser, my intention from the outset was to focus on concepts rather than details. I am no expert on life's chemical defenses—though I am not sure any one scientist is, because it is much too broad a topic. But I do have a passion for pulling together seemingly disparate ideas and a thirst for learning. This book provided many opportunities for both. While there are chapters about toxics like oxygen, metals, and ultraviolet light, and defensive proteins like metallothioneins and cytochrome P450s, they are meant as examples only and were chosen because a sufficient body of literature is currently available to outline the evolutionary history of defense. In fact, there is enough information to fill whole books about many of these topics—and so I had to pick and choose, concentrating on the most prominent or interesting literature.

I do not believe that any one of these chapter topics represents a completely new idea. Rather my hope is that by drawing connections between them, this volume will encourage students, researchers, and regulators alike to consider toxics in a broader and deeper context. Thinking about how life has evolved in response to toxic chemicals and how these systems might respond to chemicals today has been a truly fascinating endeavor. I hope that it will prove useful as well.

ACKNOWLEDGMENTS

This book covered a very broad range of topics, well beyond those with which I am proficient. For providing me with the opportunity to explore, I first thank my Island Press editor, Emily Davis. Throughout this process I have relied on the contributions of many scientists, some through their publications and others through their willingness to review chapters or answer questions. As I strove to accurately interpret and summarize the literature, their comments, corrections, and insights were invaluable. Any remaining mistakes, omissions, or misinterpretations are mine alone. The list of scholars to whom I am so indebted includes but is not limited to the following: Dula Amarasiriwardena, Caroline Bair-Anderson, Vladimir Belyi, Peter Charles Cockell, Adria Elskus, Jared Goldstone, L. Earl Gray, Mark Hahn, Amro Hamdoun, Sui Huang, Kennan Kellaris-Salenero, Michael Kinnison, Paul Klerks, Steven Letcher, Stuart Loh, Bruce McKay, Diane Nacci, Stan Rachootin, Brent Ranelli, John Saul, John Stegeman, Ann Tarrant, Gina Wesley-Hunt, and Andrew Whitehead. I would also like to thank Charles Moore and Cassandra Phillips.

I also acknowledge the contributions of the French toxicologist Andre Rico. As with any research project, my first task was to seek out previous research on the subject. An early literature search led to Dr. Rico's "Chemo-Defense System," published in 2000 and intended as a concept paper to "open up a new area of discussion and should initiate new scientific investigations."[1] Much of this paper was subsequently summarized in the EUROTOX *Newsletter*, in order to stimulate dialogue. Anticipating a lively discussion but unable to find anything, I e-mailed Dr. Rico, asking if there had been a great deal of interest in his paper. His response surprised me. Rico wrote, "I have not really had contacts concerning my papers and treated ideas. I got some good comments, not any critics. These papers were conclusions of my long experience in these areas and I am now persuaded that [there] exists for all living organisms a kind of system concerning the life in general against chemical, biological and other aggressions. I will be happy to read if you write a book on this aspect."

I acknowledge the Department of Environmental Conservation at the University of Massachusetts, Amherst, where I hold an adjunct position. Full access to the UMass library system databases and online journals made this project possible. I am also grateful to the Lady Killigrew at the Montague Bookmill and her staff, who brew great coffee and work at one of the best places to write a book.

I cannot ever adequately thank my closest friend, Penny Shockett, who has graciously read, and in some cases reread, every chapter. Her attention to detail and scientific accuracy in addition to longtime friendship has been a gift. Finally, there are not enough words of gratitude to thank my husband, Ben Letcher. Throughout our relationship, Ben has encouraged me to pursue my passions and has fully supported my desire to follow a nontraditional career path, despite the uncertainties. In reviewing these chapters, he has also offered his honest opinion, no matter the consequences.

Lastly I must thank my children, Sam and Sophie Letcher, for their patience with a mother who is home all day but has not given a thought to dinner. It is with them in mind that I continue to think more deeply about our impact on our planet's systems.

Elements

Part I: The first challenge to life's chemical defense came in the form of naturally occurring elements and other physical factors that were often both essential and toxic, like radiation

Plants and Animals

Part II: The evolution of plants and animals resulted in new complex chemicals, many of which were toxic. These new chemicals and routes of exposure further challenged life's chemical defenses

Humans

Part III: Humans have altered Earth's environments both chemically and physically. Many of these changes will put life's defenses to the ultimate test

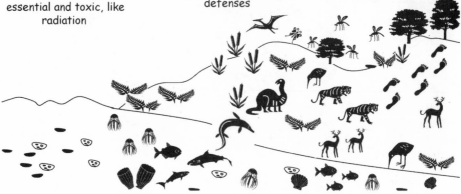

Conceptual examples: evolution of life's response to toxicants.

Chapter 1

An Introduction

The best way to envisage the situation is as follows: the environment presents challenges to living species, to which the latter may respond by adaptive genetic changes.

Theodosius Dobzhansky

All of life is chemical. But not all chemicals are compatible with life. Since their earliest origins, cells have excluded, transformed, and excreted chemicals. But sometimes a cell's defenses fail and a chemical causes damage: an organ malfunctions, a fetus is deformed, an animal dies. Toxicology is the study of these adverse effects and the protective measures that life has evolved throughout its nearly four-billion-year history. It is a science with deep evolutionary roots, and we have much to gain by better understanding the evolutionary process—whether it is how insects continually outwit pesticides, or why highly conserved metal-binding proteins interfere with the treatment of cancer. While the former, and similar cases of adaptation, have captured the attention of toxicologists and scientists interested in rapid evolutionary changes,[1] less attention has been paid to the evolution of the detoxification systems in general. For the past century, toxicologists have studied these systems, harnessing new knowledge for chemical management and regulation. We know a great deal about how any one system responds to chemicals, yet the training of toxicologists and the

1

application of toxicology seldom includes consideration of evolutionary principles.[2] Through the study of evolution, other sciences have begun to glean insights about the genesis of disease, or why some populations can consume milk and others cannot, or how wildlife management might be improved. But as ecologists, immunologists, nutritionists, and medical scientists plumb the genesis of the interactions, mechanisms, and responses relevant to their fields, toxicologists are just beginning to dip their toes in the earth's Archean waters.

Writing about the importance of turning on this "light of evolution," Theodosius Dobzhansky observed, "Without that light [biology] becomes a pile of sundry facts—some of them interesting or curious, but making no meaningful picture as a whole."[3] The word "biology" could easily be replaced with "toxicology" or any other science focused on the diversity of life and its relationship with the earth. Nearly thirty years after Dobzhansky's famous quote, an editorial in the journal *Science* proclaimed that "evolution is now widely perceived and appreciated as the organizing principle in all levels of life,"[4] while adding that the evolutionary principle is so pervasive and penetrating that it may, in a sense, be taken for granted. And we do. Although toxicologists depend on animal and cellular models, assuming common structures and functions across the broad spectrum of life, only a handful have delved into any kind of evolutionary analysis.

The toxicology of drug and chemical metabolism provides a very relevant example of how an evolutionary perspective has helped advanced the science. In the 1980s, toxicologists joked that to be published, all you needed to do was identify yet another species with a form of cytochrome P450 enzyme responsive to PCBs and dioxins (now referred to as CYP1A1). Most often the objective was to identify fish and wildlife species suitable for the monitoring of chemical contaminants. Evolution was rarely mentioned, despite the raft of papers identifying this enzyme in an astounding diversity of species, at least until the latter part of the decade.[5] We now know the CYP system is highly conserved, and this is of critical importance for understanding the evolutionary underpinnings of herbicide and insecticide resistance in plants and insects and organochlorine resistance in fish, and for predicting potentially toxic food and drug combinations in some individuals.[6]

The aim of this book then is to venture into the evolutionary history of life's response to chemical toxicants. It gathers the work of those toxicologists who have already begun looking back, and integrates their findings with relevant work by geologists, biochemists, microbiologists, physiologists, evolutionary biologists, and others. Turning the light of evolution toward toxicology, we will explore an exemplary set of defensive responses. Some, including DNA repair and antioxidants, likely appeared at the dawn of life, conserved (in most species) for more than three billion years. Others, like the p53 tumor suppressor protein, are unique to eukaryotic life. And still other protective measures blossomed only after terrestrial plant and animal life surfaced at the water's edge. Throughout this book, I refer to the network of defensive responses, for lack of a better term, as "toxic defense."[7] Revealing these responses' evolutionary roots offers a new perspective on life's ability to handle naturally occurring chemicals, as well as today's toxic synthetic and industrial chemicals.

A recent commentary explaining how physicians might incorporate evolution into medicine suggested that rather than considering the human body as the "optimally functioning" outcome of evolution, and disease as an abnormal failure, they should think of diseases as "expected and true responses to novel environmental challenges and conditions that were not present fifty thousand years ago or even fifty years ago."[8] In other words, doctors should examine how our bodies, as the products of an ancient and ongoing evolutionary process, might face new, and perhaps very different, challenges. In light of evolution, biomedical researchers are now asking questions that might seem antithetical to medicine: Has the modern-day reduction in parasite infestation and intestinal worms in many human populations led to increases in asthma, autoimmune diseases, and allergies? How useful are responses like cough, fever, and diarrhea, and when do they become a threat rather than a benefit? What is the relationship between the physiology of starvation, obesity, and diabetes?[9] "Simply put," write Randolph Nesse and coauthors in the journal *Science*, ". . . training in evolutionary thinking can help both biomedical researchers and clinicians ask useful questions that they might not otherwise pose."[10] The same could be said for researchers and practitioners of toxicology.

There is no question that we have dramatically changed much of the world's chemistry, both globally and locally. Contaminants including mercury, organochlorines, polybrominated compounds, and a host of other chemicals used in plastics, pesticides, waterproof clothing, nonstick pans, and other consumer items are now readily available to life on Earth. Looking through an evolutionary lens, toxicologists might consider how chemicals, many of them "new" to life, affect not only embryonic or fetal development, but also the development *of* the toxic response. How might such exposures influence development of a body's response to chemicals? Are there examples of comparable changes (e.g., natural yet sudden shifts in the chemical environment) in the evolutionary record? Might this help us identify responses or physiological systems most sensitive to such changes? What happens to chemicals that mimic or resemble naturally occurring chemicals—hormone mimics, for example, or nutrients? And how do we predict which chemicals will act as mimics? By considering the evolution of a body's response to harmful amounts and combinations of chemicals, toxicologists might better predict, and possibly prevent, the harm caused by today's novel challenges.

Nature's Toxicants

Throughout time, chemicals with some potential to be toxic have been both a necessity and a bane to all living things. The chemical world in which life evolved was a world where atmospheric oxygen rose from fractions of a percent to over 20%, ultraviolet light once intense and deadly now filters through a tenuous shroud of ozone, and metals, like the Cheshire cat, bounced back and forth between bioavailable and inaccessible. And these chemicals influenced not only the evolution of toxic defense but also the basic mechanisms of everyday life. There are more than one hundred known elements, which can occur in a virtually unlimited number of combinations—some naturally and some with human aid. Living things must separate the essential (or nutritional) from the nonessential while they sequester or dispose of the toxic. Sometimes, it is simply a matter of "the dose makes the poison." This has been the motto of toxicology, shorthand for the dose-response relationships that were first described by the sixteenth-century Swiss alchemist and physician Paracelsus, and it has (for better or worse) been committed to memory by new toxicologists for decades.[11] Nutrition and toxicology are often part of the same continuum, and one of life's earliest challenges may well have been maintaining nontoxic concentrations of those chemicals—essential minerals and others—necessary for basic functions. Vitamins, including A and D for example, are both necessary, yet toxic in high concentrations. And while essential metals such as zinc and copper each have their own toxic tipping point, it is plausible that the process of natural selection eventually optimized the body's response to these chemicals.[12] That is, potential harm is reduced, benefits are maximized, and trade-offs between benefits and costs are optimized. This process requires fine-tuning of all aspects of toxic defense: selective absorption, excretion, detoxification, and storage. Placing this process in an evolutionary context may provide valuable insights into a species' response to common and essential dietary chemicals *and* to chemicals that closely resemble these chemical compounds—nutritional mimics capable of bypassing exclusion and detoxification mechanisms.

Optimization of essential minerals highlights an important evolutionary principle. Evolutionary change results from a combination of environmental selection pressures. In this case, the availability of zinc influences a heritable trait, the production of a zinc-containing enzyme, and affects proteins that sequester zinc and influences their role in essential biological functions. The earth's chemical history and the changing availability of elements have dramatically influenced life's ability to defend against an overload of naturally occurring chemicals, and it may even explain why some chemicals have a greater potential for toxicity than do others.[13] The prevalence of water-soluble chemicals in seawater (carbon, nitrogen, hydrogen, oxygen, and others) at the dawn of life likely explains why some chemicals are more harmful than are others. And chemicals that were possibly more widely available before the rise in oxygen, like nickel or even cadmium, may have been used at first by early life but replaced, or displaced, as environmental conditions changed.[14] Optimization, however, cannot prepare life for major changes in environmental conditions. A useless metal may become more readily available, taking the place of an essential metal; concentrations of an essential metal may become too high; or chemicals that are relatively new to life may flood into the environment because of human activity.[15]

While not all chemicals are essential, all chemicals have the potential to cause toxicity and all living things—whether a single-celled bacterium, sea anemone, or human—must maintain chemical balance (homeostasis) in an ever-changing environment. At the very least, maintenance requires absorption of beneficial chemicals; exclusion, transformation, and excretion of harmful chemicals; and, for multicellular beings, the ability to sort vital intercellular chemical signals from the chemical noise. For complex animals that change drastically from embryo to adult and whose nutritional needs vary, maintaining balance can place different requirements on different cells and organs at different times throughout development.[16] Throughout the course of evolution, these mechanisms have been modified by reproductive strategies, life history, sex, age, co-occurring chemicals, nutritional status, temperature, the presence of certain other chemicals, and many other factors. From cell membranes to placentas, membrane pumps to complex organs, sensory neurons to brains, and single proteins to complex enzyme systems, life has evolved the ability to maintain some degree of balance. In animals that are more complex, the endocrine system, with its interconnected web of chemical messengers and receptors, is central to the maintenance of homeostasis; it is also highly susceptible to chemical-induced disruptions—a feature that toxicologists have just begun to appreciate over the past couple of decades. Yet in all species, no matter how simple or complex, the underlying cause of toxicity is the same: the defensive network becomes overwhelmed. The better we understand how the defensive network works, the better we will be able to predict when it will fail—and evolution can help us get there.

Evolutionary History of Toxicology

Before we begin our exploration, it may help to consider the other end of this equation and *its* evolution—toxicology, the ancient science of poisons and poisoning, and the modern science we rely on for protection today. We know that humans have a long history of exploiting mineral resources (e.g., zinc, lead, mercury, and arsenic) and suffering the consequences.[17] Perhaps foreshadowing our society's reliance on a host of industrial chemicals, the Romans were said to be addicted to lead. They were also aware of its darker side. The god Saturn shares his ancient symbol with lead, and "saturnine" refers to a melancholy, sullen disposition—one often associated with lead poisoning. Though Rome's aristocrats limited their own exposures, leaving the mining to slaves and the smelting to those in the provinces, they continued to drink water provided by lead-lined pipes and to sprinkle the sweet-tasting metal into their wine and on their food.[18] Some attribute the fall of Rome partially to massive lead poisoning—the first known example of large-scale harm caused by a chemical loosed from the earth's crust by humans.[19] It is also one of the first known examples of human-influenced environmental contamination.

Human reliance on metals increased both the quest to find and extract more raw materials and the incidence of illnesses associated with exposure to toxic chemicals. Some of the first documented cases of toxicity can be found in literature dating back centuries and includes the effects of lead in miners, mercury madness in hatters, silicosis in stone workers, and cancer of the scrotum in chimney sweeps. Generally, limited populations were exposed through their occupations, rather than through large-scale releases of chemicals, but observations of these exposures planted the seeds for one of the older branches of the field, occupational toxicology.

With the chemical/industrial revolution of the mid-nineteenth century came the environmental release and redistribution of historic amounts of naturally occurring and synthetic chemicals. On the heels of this chemical explosion emerged the organized science of toxicology, devoted to characterizing life's response to chemicals for the purposes of regulation, management, and exploitation. Seeking relatively quick, inexpensive, and standardized testing techniques, toxicology became a field known for its reliance on high doses, single chemicals, lethality assays, and other relatively insensitive animal-intensive techniques. This approach encouraged characterizing toxic responses as discrete or unique to one physiological system or another—the brain, the liver, or the kidney, for example—in standardized test species. Though there is no doubt we are better off today than we were even ten or twenty years ago thanks to traditional toxicity testing, there are upward of one hundred thousand industrial chemicals currently in commerce, only a small fraction of which have ever crossed the threshold of a toxicology laboratory, or have been sufficiently tested. The science of toxicity testing and its application has quickly fallen behind the chemical reality.

Over the years, advances in analytical techniques without simultaneous advances in the underlying theory of toxicology has left scientists, regulators, and managers scrambling to make sense of an ever-increasing avalanche of data. This is true even for what were once considered well-characterized chemicals. While improved sensitivity of analytical chemistry alerts us to smaller and smaller concentrations of chemicals in water, soil, blood, urine, and breast milk, molecular genetics allows us to observe altered genetic expression as tens or thousands of genes are turned on and off in response to small amounts of chemicals. And toxicologists, managers, and regulators are faced with nagging questions: What does it mean? At what point is a chemical's effect adverse? What does it mean to be exposed to parts per billion or trillion or less of chemicals like PCBs, atrazine, mercury, or plasticizers—either individually or, more realistically, in combination? And how do we interpret the reports that some chemicals typically classified as toxic in large amounts behave differently in very small amounts? Hormesis, the stimulatory response to very low doses of a chemical or physical agent, was once questioned but is now increasingly accepted as normal.[20] So at what point is the boundary crossed between an adverse effect and physiological balance, or homeostasis? Would a deeper understanding of the nature of these systems, provided through an evolutionary perspective, help to define the boundary (if one exists) between what is toxic and what is not? Would looking back into life's past help make sense of today's data?[21]

Traditional toxicology's "top-down" approach—seeking out designated end points, or worse lethality, as indicators of what is actually a highly complex response—has left too many gaps in our understanding. This is particularly true when it comes to small concentrations of chemicals or chemical mixtures. As a result, we are often at a loss when it comes to identifying and predicting the effects of today's chemical environment on living things. For example, the subtle effects of toxicants on reproduction and development were at best underappreciated and at worst unknown. When scientists revealed that a broad category of chemicals was capable of disrupting the endocrine system—resulting in behavioral changes, altered fecundity, and effects on sexual development—toxicity testing became more focused, and the number of chemicals identified as endocrine disruptors skyrocketed.[22] The spotlight on endocrine disrupters revealed both the insensitivity of current toxicological testing and the power of seeking out shared mechanisms of response. A gene turned on. A receptor activated. An embryonic barrier breached. Would an evolutionary approach—perhaps leading to insights about a receptor's selectivity—have helped toxicologists recognize the vulnerabilities of the endocrine system sooner? Could tracing the origins of the nervous system similarly improve our ability to detect and predict neurotoxic chemicals? One of the greatest challenges for toxicologists and regulators is predicting adverse effects caused by chemical mixtures. Could an evolutionary perspective help toxicologists understand how living things deal with predictable mixtures of chemicals and use this information to identify nodes in the response network that might be most susceptible to changes in this mixture?

There is no doubt that toxicology is ripe for a revolution. When the toxicologist Thomas Hartung commented in 2009 that "there is almost no other scientific field in which the core experimental protocols have remained nearly unchanged for more than forty years,"[23] he was referring to a field that we rely on every day to protect us from chemicals and exposures that have changed dramatically. We often hear about the need for a paradigm shift, or shifts that have revolutionized a particular science in the past, but we seldom have the opportunity to observe a paradigm shift as it occurs. Toxicology is at a crossroads. There may be many routes forward, but any successful path must recognize the complexity of life's responses to toxic chemicals. Deciphering the history of those responses may provide depth to a field that has too often only skimmed the surface. The revolution taking place in toxicology stands only to benefit by considering evolution.

Fortunately, our Archean history is more accessible today than ever before. Advances in chemical and biochemical techniques, including the study of genes, proteins, and their interactions on a large scale—which fall under the term "omics," as in genomics, proteomics, interactomics—provide unprecedented opportunities. Omics allow us to explore ancient origins of genes and proteins, illuminating not only the evolutionary history (or phylogeny) of cellular receptors, enzymes, and other proteins involved in the toxic response, but also the dynamics of gene and protein expression in response to contemporary exposures to foreign chemicals.[24] Already these advances are bearing fruit, laying the groundwork for evolutionary approaches to toxicology. As discussed later in this book, the combination of omics and evolutionary science has led to fascinating insights into the origins of chemical receptors and other proteins associated with toxic defense. Capturing the evolutionary history of ancient mechanisms, including the detoxification of oxygen and plant chemicals, the sequestration of metals, and the repair of radiation damage, may help toxicologists interpret and predict interspecies differences in chemical responses, understand the limitations of a receptor or enzyme, and make some sense of the highly integrated nature of these responses.[25]

Toxic Evolution in Action

Finally, though the first half of this book deals with responses evolved over billions or millions of years, sometimes change occurs much more rapidly. Referred to as "contemporary evolution" and discussed in detail in chapter 9, this phenomenon is important to keep in mind throughout the earlier chapters. Over the past two decades, toxicologists, ecologists, and conservationists have identified rapid evolutionary changes in populations exposed to various environmental stressors, including toxic chemicals. Evolutionary changes, once defined as requiring hundreds if not thousands of generations, have been observed in some species in as little as five generations.[26] We now know that evolution can occur over several decades, or a human life span, or a few centuries.[27] Once thought to be rare, particularly in vertebrates, contemporary evolution has been identified in an ever-increasing number of animals, from fish to mammals.[28] And there is ample evidence that human perturbations, including fishing and pollution, create selective pressures that induce evolutionary change in some populations and extinction in others.[29]

Knowledge of evolutionary processes, particularly contemporary evolution, can profoundly influence the management of populations affected by human endeavors—whether fishing, clear-cutting, or the release of toxic chemicals.[30] Yet acting based on observations of contemporary evolution alone would be like beginning at the end. Without some knowledge of more distant evolutionary history, predictions based purely on observations of rapid evolution could mislead.[31] While omics may reveal the places we have been, studies of evolutionary history and contemporary evolution could tell us how we got here.

Moving Forward by Looking Back

If we had not changed the world through our indiscriminate use of elemental and industrial chemicals, we would have little use for toxicology or the evolutionary history of toxic defense. Yet we have, and the consequence is that we have introduced unprecedented opportunity both for contemporary evolution and for evolutionary mismatches. These mismatches occur when environmental change results in conditions divergent from those in which a species or population evolved. Writing about the inevitability of mismatches, the geneticist Sean B. Carroll commented, "Evolution and the DNA record of life tell us that natural selection acts only on what is useful for the moment. It cannot preserve what is no longer used, and it cannot predict what will be needed in the future. Living for the moment has the dangerous disadvantage that if circumstances change more rapidly than adaptations can arise, faster than the fittest can be made, populations and species are at risk."[32] Understanding the evolutionary history of a particular toxic defense mechanism may reveal a number of important character- istics of defense, including past selective pressures, phenotypic plastic- ity, rates of evolution under "natural" chemical conditions, and link- ages to other traits involved in chemical responses. Maybe someday this depth of understanding will become a routine part of toxicology. It is an exciting endeavor, yet one that cannot possibly be fulfilled by a single book or by a single author. Instead, my goal is to open the door a crack, and peer into the fascinating history of life's response to toxic chemicals. This book is intended to be just one small step back into the abyss of time.

I have tried to organize topics when possible in chronological or- der. Where the evidence allows, linkages between chapters and systems are made. Yet each toxic challenge and defense mechanism is intended as an introduction to a broader concept—the ancient and highly con- served nature of life's chemical defense mechanisms—rather than as a detailed literature review of any one of life's protections. As we travel forward in time, we first explore the roles of the earth's chemical and physical elements on evolution, from the ultraviolet radiation on DNA repair (chapter 2) to the emergence of enzymes that protected our ear- liest ancestors as highly toxic oxygen flooded the planet (chapter 3). Like ultraviolet light and oxygen, metals are one of life's oldest chemi- cal challenges, and one solution was the evolution of metal-binding proteins, as explored in chapter 4. These three chapters make up the "Element" section of this book, and provide brief introductions to a select set of defense mechanisms evolved in our earliest single-celled ancestors. In the following section, "Plant and Animal," we move from single- to multicellular animals, and explore defenses evolved to pro- tect these more complex beings (some originating in single-celled life, only to blossom in their multicellular descendants). Chapter 5 exam- ines the battle between cells, the emergence of cancer (a disease spe- cific to multicellular organisms), and the evolution of p53, the anti- cancer gene. As life moved from an aquatic existence to a terrestrial one, plants and animals evolved their own toxic biomolecules to foil predators, and predators evolved enzymes to detoxify them. This re- sulted in ongoing plant-animal warfare and a highly effective detoxifi- cation system (chapter 6). As mentioned earlier, chemical receptors are an integral part of detoxification and of chemical sensing. Chapter 7 explores the evolutionary history of two different receptor systems. Finally, the networked nature of all these systems, and many others, is introduced through a discussion of environmental response genes in chapter 8. The last section, "Human," comprises only two chapters. The first (chapter 9) addresses our role in facilitating contemporary evolution, while the last chapter (chapter 10) contemplates the overall impact of industrial and synthetic chemicals in light of evolution. Humans have transformed the world in many ways—including altering the availability of myriad chemicals. Though we logically focus on the adverse effects of modern synthetic and industrial chemicals, life's protective systems are ancient. In this rapidly changing world, looking back may actually allow us to move forward, and so we begin at the beginning.

PART 1

Element

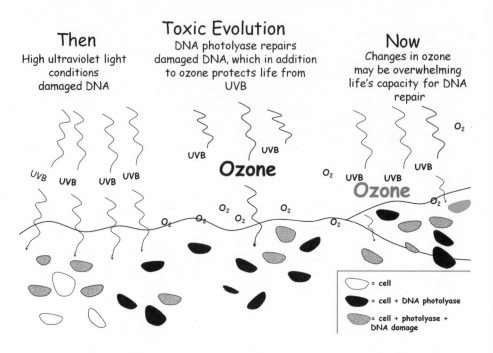

Ultraviolet light and the evolution of DNA photolyase.

Chapter 2

Shining a Light on Earth's Oldest Toxic Threat?

To humans, the Earth is a photobiologically protected haven encircled by a fragile O_3 shield that has been perturbed by the activity of industry. This view of the recent perturbation of the O_3 column is certainly accurate, but it is clear that the Earth has been subjected to quite varied UV regimens throughout history.

Charles S. Cockell

Between 3.4 and 3.8 billion years ago, life happened. While scientists debate whether life first emerged on the planet's surface in an organic "primordial soup,"[1] was helped along by meteorites,[2] or evolved in the ocean's black smokers,[3] there is little doubt that it developed on a planet flooded with massive amounts of ultraviolet radiation (UVR). UVR is a highly energetic and destructive force that may have actually aided evolution. UVR-induced mutations induced in deoxyribonucleic acid (DNA) struck at the very heart of life, and when not lethal, may have accelerated the pace of change. (UVR may even have played a key role in the formation of RNA nucleotides, possible precursors to DNA in the earliest stages of life's formation.)[4] In either case, high mutation rates would have resulted in precarious conditions requiring protection if life was to evolve into the complex forms known today. In a highly coordinated process essential for both reproduction and survival of all living things (and viruses), DNA bonds break and

15

reform, transferring, capturing, and storing chemical energy. Owing to their molecular structure, nucleotides like DNA and RNA are chromophores. That is, their chemical nature causes them to readily absorb UV light (more than, say, proteins), and so they are particularly susceptible to UVR, which, in sufficient amounts, breaks DNA's bonds and modifies life's genetic code.[5] If not repaired, these UVR-induced changes cause permanent mutations or death. Today we rely on the earth's stratospheric ozone layer to protect us from excessive UVR, but more than three billion years ago, when life first emerged, there was no ozone layer. The threat posed by UVR was far more hazardous then than it is now. This potentially lethal interaction between life's basic genetic code and UVR makes it one of the first known physical toxicants still relevant today. As such, it is a good place to begin our exploration of the evolution of toxic defense.

The First Threat?

As a toxicologist who was far more interested in biology than chemistry or physics, I've reluctantly accepted the reality that in order to fully understand the biology of toxic defense, one must also understand the chemical or physical threat—in this case, sunlight. Whether UV or visible, the sun's light constitutes a small portion of the electromagnetic energy spectrum, which ranges from radio waves, with wavelengths spanning hundreds of meters, to X rays and gamma rays, mere trillionths of a meter in length. While "visible" light waves range from 390 to 750 nanometers (nm), UVR inhabits the lowest end of the light spectrum, beyond our capacity for visible detection. Ultraviolet radiation is split into three relatively arbitrary ranges. Ultraviolet A (UVA), from 320 to 400 nm, may be familiar to those of us of a certain age who lit up our rooms under blacklights, or those who continue to visit the tanning salon despite the risks. After years of seeking a "healthy tan," we now know that in addition to tanning, UVA causes skin aging and skin cancers; the recent accumulation of data has prompted the U.S. FDA to revisit its current regulations and recommendations for UVA exposures.[6] In contrast, ultraviolet B (UVB) (290–320 nm) is notorious for its role in causing sunburn, skin cancer, and immunosuppression (yet the role of UVB in the production of vitamin D also means that some UV light is essential for most life forms).[7] Finally, there is ultraviolet C (UVC), which ranges from 200 to 290 nm.[8] Often used for industrial sterilization, UVC is the most highly energetic and destructive UVR.

So, with all this damaging radiation, why isn't skin cancer even more prevalent than it is today? How do frog eggs survive when exposed to the beating sun? And why isn't UVR a problem for species that depend on basking in the sun to warm up their bodies? All this can happen, in large part, because of the stratospheric ozone layer. We might have paid little attention to the ozone layer or its destruction were it not for its role in absorbing most of the UVC and a large portion of UVB. Because of stratospheric ozone, over 90% of the UVR reaching the earth is UVA. But that wasn't always the case. Four billion years ago, young Earth was lacking not only ozone, but also another important protection against emissions from the sun: a large-scale magnetic field. Recent studies suggest that establishment of the magnetic field, around 3.45 billion years ago, might not only have predated life, but also may have also provided the planet with its first line of protection against the sun's high-energy ionizing radiation (an even more destructive form than UVR), allowing for the emergence of life as we know it.[9]

While the magnetic field may have been a prerequisite for life, life was a prerequisite for the formation of the next major line of defense, stratospheric ozone. Ozone is produced through interactions between oxygen gas and UVR. Yet the nearly 20% oxygenic atmosphere we enjoy today was nonexistent on early Earth. So where did all the oxygen come from and when did it begin to accumulate? Like the origins of life, there is no simple answer—and even attempting to sort through all the theories and hypotheses is well beyond the scope of this book.[10] Because of its role in forming the ozone layer, however, a few words are in order. How much oxygen was available when and by what process is an active area of research, yet there is some agreement that well before the production of oxygen by photosynthetic species, very small amounts of O_2 could have been produced by interactions between UVR and water vapor. At the time life is thought to have emerged, O_2 concentrations are estimated to have been less than one thousandth of today's levels.[11] But rather than building up, those small amounts of oxygen gas are thought to have reacted with available iron and other reducing agents, tucking it away in the earth's crust. There is also some agreement that the amount of oxygen gas required to form an ozone layer of appreciable merit did not exist before the evolution of adequate numbers of photosynthesizers—the earth's primary producers of oxygen gas. Some have hypothesized that photosynthetic organisms appeared very early in life's history, well before buildup of large percentages of atmospheric oxygen.[12] Although their existence is relevant to the discussion of oxygen toxicity (chapter 3), it is thought that any oxygen they may have produced, like that formed by UVR, would have combined with easily oxidized minerals rather than forming stratospheric ozone. Consequently, by some estimates, an effective ozone shield did not exist until photosynthetic life was in full bloom, roughly 2.2–2.5 billion years ago.[13] Earth's surface, including shallow water environments, would have been a harsh place for life before photosynthesis.

The geomicrobiologist Charles Cockell surmises that on a planet devoid of ozone, life survived a barrage of UV light that was hundreds or even thousands of times more powerful than it is today (when "weighted" based on DNA damage associated with UV wavelengths), despite the reduced luminosity of the earth's young sun.[14] It was a planet where the near surface of the Archean ocean would have been far less welcoming, and maybe even outright lethal, compared to the uppermost regions of today's marine habitats. At depths of up to five meters below the water's surface, Cockell estimates that living cells would have contended with UVR intensities one hundred times more likely to cause DNA damage than the intensities on the ocean's surface today.[15] So whether life emerged in the ocean's black smokers, far from the reaches of UV light, in a primordial soup energized by UV light, or by some other means, at *some point* primitive life not only survived but also reproduced under high-intensity UVR. The continual survival and further evolution of life, particularly before the buildup of the ozone layer, would therefore depend on its capacity for protection, avoidance, or repair of UVR-induced DNA damage.

While stratospheric ozone blocks the sun's harshest rays, enough UVR reaches the planet's surface that without additional protection, higher life forms (as we know them) would likely not exist today. UVR is the single most important risk factor for some skin cancers in humans, and remains a threat to a broad range of life, from fish eggs congregating in the ocean's uppermost microlayer to coral reefs to Cascade Mountain frogs. Living things rely on a repertoire of protective, repair, and avoidance responses—some of which, not surprisingly, can be traced back more than three billion years.

Over time, life evolved complex dependencies on the sun—requiring both exposure and protection. Photosynthesizers require visible sunlight to produce food, and we in turn depend on them for food, oxygen, and sopping up CO_2. Humans and other species, from phytoplankton to frogs, rely on UVB for vitamin D production. And it is likely that there are many other requirements for UVR we have yet to discover. Even as early life avoided UVR by sheltering in muck, under rocks, in the deep ocean, or beneath layers of dead cells, when our earliest ancestors began living more directly under the sunlight, they needed a way to prevent or repair the inevitable DNA damage. Pigments capable of absorbing and diffusing UV energy, for example, may have provided a natural sunblock, while enzymes repaired DNA damage from errant UVR. Better than rocks or sediment, these protections were internal—capable of protecting primitive cells wherever they drifted, floated, or roamed. Given the essential nature of DNA and the threat posed by sunlight, it should come as no surprise that a DNA-repair enzyme specializing in damage caused by UVB, better known as DNA photolyase, is among the oldest enzymes known to life.

The DNA molecule is both ancient and surprisingly simple, consisting of double strands of sugar-phosphate backbones beaded with pyrimidine (thymine or cytosine) or purine (adenine or guanine) nitrogenous bases. These four bases, with purines linking up with their pyrimidine partners on a complementary strand of intertwined DNA, make up life's genetic code—a simple but elegant code for an amazing diversity of complex life. Pyrimidine bases are also the targets of UVB. The most common lesion caused by a cell's absorption of UVR[16] is the abnormal bonding of adjacent pyrimidines (e.g., two thymines), a process called dimerization, caused by UVB. This bonding distorts the double helix, causing kinking of DNA.[17] The combination of pyrimidine dimers and DNA kinking in turn causes lesions (and eventually mutation), which if not repaired could adversely affect reproduction, cause cell death, or, in the rare event, provide the raw genetic material for evolution. Some regions of DNA, those with particular base combinations that confer greater flexibility, may be more susceptible to this kind of damage than are others;[18] researchers are just beginning to understand the role of mutation-prone regions of DNA in evolution.[19] Flexible regions are also important for gene regulation—such that UVR-induced damage could be quite detrimental.

Enter DNA cyclobutane pyrimidine dimer photolyase (or DNA photolyase), with a single known role—cleaving pyrimidine dimer bonds and returning DNA to its normal functioning state.[20] Ironically, photolyase belongs to a rare class of enzymes that require sunlight (in this case, blue light ranging from 320 to 500 nm) to undo the sun's damage. As the photolyase binds to a dimer, the sun's energy breaks the dimer bond and reverses the damage. One could imagine that because DNA was susceptible to mutation by UVB, and since UVB was so prevalent, as our ancient ancestors settled into a life under the sun, repair by enzymes like DNA photolyase became a necessity.[21] So how ancient is DNA photolyase and what of its role today, given human-induced changes in the stratospheric ozone layer?

Genealogy of Protection

As scientists hypothesize about the influence of early Earth conditions on the origins of life or the production of oxygen, powerful new techniques such as genomics and bioinformatics are providing insights into the earliest occurrence and evolution of key enzymes and proteins like DNA photolyase. Genomics refers to the sequencing and study of an organism's entire complement of DNA. This includes genes coding for structural and functional proteins (like DNA photolyase), and, depending on the species, what for now appears to be copious amounts of "noncoding" DNA.[22] Through genomics, we now know that the number of genes in an organism (other than viruses) can range from several hundred in some archaebacterium and bacteria to tens of thousands in some plants and animals. Human genes, for example, number a bit over twenty thousand. This is well below the one hundred thousand originally envisioned, and surprisingly similar to the gene totals estimated for our very distant cousin, the sea urchin.[23] Genomics has opened the genetic floodgates, yet an evolutionary geneticist, toxicologist, or biologist wishing to trace the origins of particular genes could get lost in the tens of thousands of genes and the even greater number of DNA nucleotide base pairs that make up each gene—if not for bioinformatics. The successful marriage of computer science and information technology—bioinformatics—allows scientists to make sense of large data sets, like the functional genomes of several dozen or more species. These advances will allow us to peer into the past not only of DNA photolyase, but also of a number of proteins and enzymes critical for defense against potentially harmful chemicals. Yet there is always the caveat that with additional information or analytical techniques, conclusions today are subject to modification tomorrow.

Ideally, if we wished to trace the origins and evolution of DNA photolyase genes using genomics, we would simply seek out the living relatives of the most ancient organism, and compare the photolyase gene sequence with those from a range of current, or extant, species, using an appropriate bioinformatics analytical method. This is, of course, a toxicologist's or evolutionary biologist's fantasy. As alluded to earlier, there is little agreement about the most basic characteristics of the tree of life's deepest roots, let alone which organism living today is most closely related to the earliest life on Earth. Was this ancestor complex or simple? Anaerobic or oxygen-tolerant? Did it form in the deep ocean or in a prebiotic soup? And is a tree even the most appropriate analogy?[24] *Where* life evolved is relevant to our quest and raises interesting questions about enzymes like DNA photolyase. If life evolved in the ocean's black smokers, why would it have benefited from DNA photolyase (assuming UVB damage repair is its primary function)? Or, if instead life evolved nearer to the ocean surface, which came first, DNA photolyase or the capacity to *avoid* the sun's damaging emissions? Or were they coincident? Unfortunately, until the prebiotic dust settles, these answers remain beyond our grasp.

Fortunately we need not wait for confirmation of the *very first* appearance of the enzyme if our goal is to consider how its presence in extant species and its evolutionary history might help us better understand the impact of UVR on life today. Instead, we might focus on how it came to be distributed in the great majority of extant species—and then seek out interspecies differences in the enzyme's evolution. Was there a common origin for DNA photolyase? Or are DNA photolyase enzymes examples of convergent evolution—functionally similar yet structurally different, indicating a different means to the same end? Are there some species living without the enzyme, and did they lose it or just never have it? And are those species more susceptible to UVB? For clues we might move along to one of the next intriguing, yet no less controversial, junctions in life's evolutionary course: the last universal common ancestor, better known as LUCA. Although LUCA's most basic characteristics remain unknown (was LUCA a prokaryote or a eukaryote, or was it more appropriately envisaged as an "ancestral state" rather than any one particular entity?),[25] there is much to be gleaned from what is known. In 2006, Christos Ouzounis, the computational biologist co-credited with "baptizing" the term *LUCA*,[26] shed some light by reconstructing LUCA's hypothetical genome. Analyzing 184 sequenced genomes, Ouzounis and colleagues estimate that LUCA's "minimal gene content" hovers around 1,500 gene families.[27] While the results may not answer questions regarding LUCA's lifestyle, based on the known functions of the majority of these gene families, they do suggest that our common ancestor had already developed a range of functionality—not unlike many extant species.[28] Most relevant for us is that today, several of these gene families are associated with chemical defense and repair. Number 231 in Ouzounis's list of functional genes is deoxyribodipyrimidine photolyase, or DNA photolyase. This finding confirms, at the very least, the ancient origins of this enzyme. While the primary function for this (or any) gene may change over time, the simple function of DNA photolyase suggests that at least in this case the enzyme's role in DNA repair has been conserved for a very long time. Not only is DNA photolyase truly ancient, but like other enzyme systems discussed in subsequent chapters, it remains widespread today, occurring in organisms ranging from bacteria to invertebrates and vertebrates. Since repair by DNA photolyase is both rapid and efficient, its highly conserved nature should be of no surprise.

Yet, active DNA photolyase is missing from at least one branch of the family tree—placental mammals, including humans.[29] This loss of photolyase has been tracked back roughly 170 million years,[30] to a time when placental mammals may have first parted ways with their marsupial cousins (the branch of the mammalian family that retained DNA photolyase). Perhaps losing a solar-powered gene that corrects for solar damage was of little consequence for a class of vertebrates who likely evolved as nocturnal creatures, and whose embryos were well protected within a placenta rather than exposed to the elements. But what sets placental mammals apart from marsupials? While there are no clear explanations, some researchers speculating on gene loss in bacteria living in low-UV environments[31] have suggested that in placental mammals, the combination of reduced "need," in addition to the possible presence of a redundant albeit less efficient repair system (discussed below), may have allowed for the loss or alteration of a functional DNA lyase.[32]

In an interesting aside, cryptochrome, a protein involved in maintaining the circadian clock in mammals, is thought to have derived from the DNA photolyase gene following a gene duplication event. Although cryptochrome itself may be off topic (although linkages between daily rhythms and protection from the sun could certainly be argued as relevant), the process of gene duplication is worth a few words because it will appear many times throughout this book. Sometimes a single gene, a region of a chromosome, or a whole chromosome is duplicated through an error in replication. This gene is redundant—presumably there is still a properly functioning copy. But rather than necessarily harming the organism, the duplicated gene is freed from selective pressures, resulting in a greater potential for accumulation of mutations, changing the form and function of associated proteins. Duplication, in addition to mutation of functional genes, serves as a basis for evolution.

Returning to the mystery of our missing DNA photolyase, does the "reduced" need theory fit? *Traditional* evolutionary theory holds that evolution generally proceeds through preservation of beneficial changes or mutations and elimination of harmful changes, suggesting that perhaps, at least for a time, DNA photolyase may have been more costly than beneficial. Subsequent modifications to the theory propose that rather than relying primarily on "positive" selection for adaptive mutations, natural selection may instead eliminate harmful mutations, while "genetic drift" (the change in gene frequency as a result of chance in small populations) provides a means of fixing neutral mutations.[33] Many species, including mammals, have many different DNA repair systems—or redundant systems—albeit some are more efficient than are others. So it is possible that a useful and efficient gene could possibly be lost if other genes or groups of genes were able to cover the same functions. A recent analysis of photolyase gene loss by the evolutionary geneticists Jose Lucas-Lledó and Michael Lynch suggests that eukaryotes may have simply lost certain gene functions through a buildup of non-adaptive mutations—which, combined with an inefficient selection process and a functional backup, allowed for perpetuation of potentially ineffective genes throughout a population.[34] Organisms with relatively small population sizes (e.g., mammals in comparison to bacteria), hypothesize Lucas-Lledó and Lynch, may be more prone to inefficiencies in natural selection, such that "the complete loss of photolyase activity in many eukaryotic lineages, including placental mammals, may not be adaptive."[35] Or it may not be the result of adaptive processes, an intriguing idea that may shed some light on other mechanisms that are protective yet less than ideal.

So what is this less-efficient but adequate backup system on which we depend for protection from UVB? Recall that the primary damage caused by UVB light is the bonding of two pyrimidine bases along DNA's backbone, distorting the DNA helix, which must then be repaired for normal DNA replication and function to proceed. While not as elegant as photolyase, the nucleotide excision repair pathway, or NER, is the primary, maybe even sole, repair system in humans and other placental mammals for this DNA damage.[36] Also referred to as "dark repair" because it doesn't require light, the NER system is a complex system involving dozens of different steps—from detection to excision to repair—each performed by dozens of different gene products.[37] Though not the most efficient system in terms of resources and possibly speed, NER is quite important for the repair of UV-induced damage in humans, as highlighted in individuals with xeroderma pigmentosa (XP), an inherited disorder involving the NER. The disease, named and described in 1882 by Moritz Kaposi, refers to hypersensitivity to the sun, premature skin aging, and a significantly increased risk of skin cancers.[38] It can be traced directly to improper functioning of key genes in the NER pathway. The evolution of a complicated system requiring dozens of enzymes to cover the same function as a single lost enzyme is intriguing. While we may have an explanation for enzyme or gene loss (e.g., inefficient natural selection, a nocturnal lifestyle), by what mechanism might a far more roundabout way take its place? Did the NER pathway evolve de novo, eventually filling the void left by the loss of photolyase? Or did it fortuitously emerge from the smorgasbord of existing DNA repair mechanisms, some of which also originate as far back as LUCA?[39]

No matter the route, it is clear that repair of UV-induced DNA damage is a process that is both highly conserved and that, in some species, continues to evolve. One explanation for the evolution of the NER pathway in placental mammals suggests efficiency: most of the genes already existed and were simply co-opted for UVB repair. In their review, titled "Quality Control by DNA Repair," Thomas Lindahl and Thomas Wood observe, "One fascinating feature of mammalian NER proteins is that most of them have dual functions, participating in other aspects of DNA metabolism."[40] In fact, one of the core genes, XPA, which codes for a protein that binds to damaged DNA and helps organize other NER proteins, is one of the few that has no other known function. A mutation in the XPA gene also causes XP. If XPA somehow "rounds up" dual-function proteins and directs them toward NER, it would suggest that evolution of only one or two new gene products may have been necessary to replace photolyase. On the other hand, some studies suggest that rather than a highly conserved network of proteins, the NER has evolved several times over in prokaryotes and eukaryotes through convergent evolution.[41]

So why is UVB still a problem? Given the long evolutionary history, and the general trend (until modern times) toward reduced UVR, the potential for UVR-induced harm should be of little concern to toxicologists, environmental managers, and regulators. Unfortunately, between climate change and the thinning of the ozone layer, alterations in the physical environment have, for at least the second time in evolutionary history, altered the relationship between the sun's UV light and life on Earth. When environmental conditions are rapidly and vastly changed from those in which a species evolved—increased UVB in this case—it presents an evolutionary mismatch. The present-day changes have created a potential evolutionary mismatch on a global scale—and the effects are beginning to show.

Everything under the Sun

It was under the protection of stratospheric ozone that life could more broadly colonize the earth, an atmospheric condition that humans altered in a blink in time. Over a period of a couple hundred million years, the ozone layer became and remained relatively stable, with predictable seasonal changes and occasional large but short-lived fluxes caused by catastrophic events—a volcanic explosion, an asteroid, or a solar flare.[42] Fast-forward to the early 1980s, when the British scientists Joseph Farman, Brian Gardiner, and Jonathan Shanklin first reported a 10% decrease in the ozone layer hovering over the Antarctic.[43] Their finding marked the first known large-scale, human-induced change in stratospheric ozone. Subsequent studies revealed an enlarged ozone hole appearing each spring as far back as the mid-1970s, and that the hole was growing. The consequence of ozone depletion? Over half a century of increased terrestrial levels of UVR worldwide, which will likely occur for decades to come. Several years before the ozone hole was discovered, the eventual Nobel Prize winners Frank Rowland and Mario Molina reported that certain halogenated chemicals could cause chlorine to build up in the stratosphere. This accumulation, in the presence of ultraviolet light and cold temperatures, destroys ozone. And unfortunately, the responsible chemicals, chlorofluorocarbons (CFCs), were commonly used in a multitude of consumer products—from hairspray to freezers. While this research eventually led to the first ban on CFCs in the United States, it applied only to aerosols, thanks to industry push back. The role of

CFCs in creating the ozone hole, first observed by Farman and colleagues, was confirmed through the work of the atmospheric scientist Susan Solomon. This paved the way for the 1987 UN Montreal Protocol on Substances That Deplete the Ozone Layer, the international agreement protecting the ozone layer through the regulation of ozone-depleting chemicals.[44] A 1991 follow-up assessment on the environmental impact of ozone depletion described the effects of increased UVB on skin cancer and immunosuppression in humans, and on marine phytoplankton and other species already under UVB-induced stress.[45]

Though we have reduced ozone depletion in the most severely afflicted Antarctic regions, the thinning continues, but at a slower pace. World use of CFCs has dropped from over 1 million ozone-depleting (ODP) metric tons before the Montreal Protocol, to roughly 1,000 ODP metric tons today.[46] But the effected regions are now known to extend far beyond the Antarctic and into the temperate zones.[47] In his 1995 Nobel acceptance speech, Sherman Rowland talked about the role of ozone depletion on UV radiation:

> On October 26, 1993, when a particularly low ozone value existed over Palmer in the Antarctic peninsula, the UV-B intensity exceeded by 25% the highest intensity recorded in San Diego, California, for any day of 1993. . . . The remarkable further observation was also recorded that the most intense weekly UV-B exposure in any single week at any of these stations was recorded at the South Pole. There, even though the sun is low on the horizon, the very low concentrations of ozone in the late Antarctic spring, the general absence of clouds, and the 24 hours per day of continuous sunlight combined to permit a higher weekly dose of UV-B than in any week in San Diego, California. The often-heard statement that UV-B intensities in the Antarctic can never be very large is no longer true. The question of possible biological damage associated with UV-B radiation then requires the separate assessment for each biological species of whether such damage is the result of cumulative UV-B exposure over an entire season or of a single, extremely intense exposure.[48]

Considering Rowland's comments, how might species like krill and other Antarctic life forms fare when the ambient UVB starts to resemble that of San Diego, California? Do they suffer sunburn? Increased

DNA damage? Or are their DNA repair and protection mechanisms up to the task? And what about more temperate species, like frogs and other amphibians, living at high altitudes where UVB is naturally higher: Would increased UVB exposure affect them? Has increased UVB contributed to the decline of frog populations around the globe? And are there additional causes of increased UVB exposure that we should consider in addition to ozone depletion—like climate change–induced habitat and physiological changes? More than twenty years after the discovery of ozone thinning, scientists now know more about the effects of increased UVB on certain species, most notably amphibians, and have begun to evaluate these effects in the context of at least one of life's conserved methods of UVB protection, DNA photolyase.

A Plague on Frogs

Years ago, while hiking the Oregon Cascades, my husband and I pondered the fate of thousands of tadpoles that filled ephemeral ponds and puddles along our route. At the time, we'd wondered how they'd manage in their rapidly shrinking aquatic environment. Would they metamorphose and climb onto land, or shrivel up without a chance in the high mountain sun? As we slathered on the sunscreen, it never occurred to us to consider how these tadpoles contended with the increasing risk of UV light as their protective aquatic environment dried up. Perhaps we didn't think much about it because we figured they were obviously well adapted for their environment—otherwise they wouldn't be there.

Since then, dramatic declines in amphibian populations[49] spurred research into the effects of a host of human impacts, including climate change, pesticides and herbicides, disease, and UVB. Because amphibians have little in the way of protective scales, shells, fur, or feathers, and because they often lay their eggs in water, they are particularly vulnerable to environmental conditions. Regardless of whether it is the cause of large-scale population declines, it is now clear that exposure to UVB radiation causes a range of effects in amphibians, from reduced hatching to developmental malformations and greater susceptibility to pathogens—depending on the species, and sometimes even the specific populations of a single species.[50]

Andrew Blaustein, an ecologist at the University of Oregon, has studied frogs for decades, and for the past ten years he has turned his

attention to the role of UVR in population declines.[51] Like many ec-
totherms (animals formerly known as cold-blooded), some frog species lay their eggs in sunlit ponds or puddles, expressly relying on the
sun's energy to speed along egg hatching, larval development, and
metamorphosis before their ephemeral pond dries. Much like photosynthesis or vitamin D production, it's a trade-off—in the frog's case,
faster development in a higher-risk environment. Of course, like most
creatures living under the sun, amphibians are well defended against
UVB radiation. In addition to behavioral changes, like burrowing in
mud or laying eggs in logs in under rocks, and the production of natural sunscreens, amphibians have redundant systems for DNA repair,
including DNA photolyase.

Interested in the level of protection afforded by DNA photolyase,
and the potential impacts of increased UVB exposures on frog populations, Blaustein and coauthor Lisa Belden compared the life history
habits of several amphibian species with DNA photolyase activity in
their eggs. Their study reveals strong positive correlations between
UVB-resistant frog species (a species, for example, whose eggs are
normally most exposed to sunlight because they are laid in sunny shallow ponds) and increased photolyase activity, in comparison to species
whose eggs tend to be protected from direct sunlight. In other words,
frog species that lay their eggs in sun-drenched environments are better able to repair DNA damage caused by UVB. Not only that, but
subsequent field studies confirmed the detrimental effects of naturally
occurring levels of UVB to developing eggs of some frog species,
while those with the highest concentration of photolyase, the Pacific
tree frogs, were most resistant.[52] Beyond killing embryos, write Blaustein and Belden, UVB exposure may also cause sublethal and potentially subtle (and therefore more difficult to measure) effects on larval
growth and development. Their findings raise a couple of intriguing
questions. Are less-resistant species more susceptible to DNA damage
caused by increased UVB? And does UVB interact with other environmental stressors (either naturally occurring or human induced)?

The first question was answered in part by researchers working
with a single species of frogs inhabiting different altitudes of the
French Alps. Frog populations adapted to life at higher altitudes, and
therefore naturally higher UVB exposures, showed less DNA damage
than did their lower-altitude brethren when exposed to UVB intensities typical of high altitude.[53] Identifying the genetic mechanism
of this adaption—rapid evolution, increased protein production, or

both—will require further study. Although DNA photolyase concentrations were not measured, the authors report an interesting twist that suggests increased photolyase activity in high-altitude tadpoles. Interested in other ways frogs might experience DNA damage, they studied the effects of benzo(a)pyrene (BaP), a well-characterized carcinogen present in cigarette smoke, coal tar, oil, and myriad other combustion products. BaP is both an ancient toxicant and a major industrial pollutant. Activated BaP (technically the chemical is a procarcinogen, which must be activated to its carcinogenic form, most often through metabolism) binds with DNA, causing a kink in the DNA helix, just like UVB. Recall that a specialty of DNA photolyase is kinky DNA. It turns out that high-altitude frogs had less BaP-induced DNA damage compared with their lowland cousins.[54] Added protection by DNA photolyase? Maybe. Until enzyme concentrations are confirmed, any added protection cannot yet be attributed to increased DNA photolyase.

Although separate from the DNA photolyase story, there is one more study of interest that addresses the question of combined exposures to environmental stressors. An investigation by Joseph Klesecker along with Blaustein and Belden[55] suggests that frogs exposed to higher UVB may also be more susceptible to infection by the fungus *Saprolegnia ferax* (as mentioned earlier, UVB also causes immunosuppression). In this field study, the authors altered UVB exposure by manipulating water levels in mountain breeding ponds—UVB intensity increased as water depth decreased. The authors concluded that frog populations may be exposed to higher levels of UVB because of both disappearing ozone and ponds made shallower by climate change—a potential double whammy. More recent studies, however, point to a newly identified chytrid fungus as a major player in worldwide frog declines.[56] While immunologists focus on the ability of susceptible species to mount an immune response, the role of UVB in frog declines and in altered immune response in frogs remains unclear. Even if UVB is eventually exonerated in the case of amphibian declines, the set of studies on UVB and frog eggs is a good demonstration of both the reliance of high-altitude frogs on DNA photolyase for protection from UVB, and the potential for this ancient defense mechanism to be overwhelmed by changing environmental conditions. It also begs the question, how might other species exposed to increasing UVB intensity (fish eggs in the ocean's microlayer, for example) fare in today's world?

Cosmic Irony

Over time, populations strike a tenuous balance between survival and combating naturally occurring physical toxicants. In this case, the production of DNA photolyase must be balanced with all the other energetic needs involved in the maintenance and reproduction of life. We might have paid little attention to the photolyase enzyme and its role in repairing UVB-induced damage if we had not upset this balance. The side effect of our reliance on halogenated chemicals to combat the sun's warmth—reduced stratospheric ozone—resulted in more UVB reaching the planet's surface. We are just now beginning to understand the consequences. Macroevolutionary changes, such as the loss of DNA photolyase in placental mammals, and microevolutionary changes in populations of UVB-resistant and -sensitive frogs, reveal the complex relationship between life and its surrounding environment. These changes should teach us something not only about life's capacity for resilience, but also about humans' capacity to overwhelm millions, maybe even billions, of years of evolutionary adaptation. Fortunately, despite laying the groundwork for a catastrophic evolutionary mismatch, we have recognized our folly and changed our ways by reducing and replacing harmful chemicals. As the ozone recovers, we can study the consequences of pushing one of life's oldest chemical defense mechanisms to its limit. If we continue to learn, by the time atmospheric ozone returns to pre-CFC levels, we may be better able to predict and prevent other destructive environmental changes.

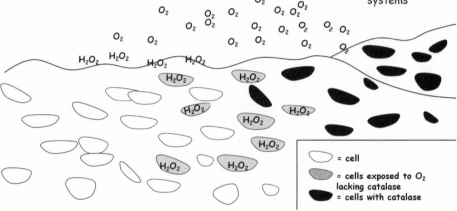

Then
Low or no oxygen meant an anaerobic lifestyle

Toxic Evolution
With increased oxygen life was challenged with reactive oxygen species. At least one enzyme, catalase, was available to detoxify the ensuing, reactive, hydrogen peroxide

Now
Although humans have yet to alter atmospheric oxygen we constantly challenge our ROS detoxification systems

= cell

= cells exposed to O_2 lacking catalase

= cells with catalase

Oxygen and the evolution of catalase.

Chapter 3

When Life Gives You Oxygen, Respire

Despite millions of years of evolution, we still cannot adapt to the concentration of oxygen that nature provided us.

Nick Lane

Oxygen: we cannot live without it, yet every day we struggle to coexist with this highly reactive and potentially toxic chemical. While the evolutionary history of the earth's atmospheric and oceanic oxygen is a controversial topic for geologists and evolutionary biologists alike, there is no question that oxygen, like UVR, requires detoxification even in the smallest amounts. Yet it is also essential for life as we know it. Survival in an oxygenic atmosphere required the evolution of detoxification responses, and recent genomic analysis is helping to reveal the ancient origins of these defense systems. While theories about the earth's historical oxygen concentrations are constantly in flux—ranging from a young planet with occasional pockets or "whiffs" of oxygen, to a virtually anoxic planet where anaerobic life flourished for billions of years—at some point, oxygen began to accumulate in the atmosphere, and life forms evolved that could not only cope with, but also exploit the highly reactive gas. Our ability to breathe deeply today has deep evolutionary roots.

As with the previous chapter on ultraviolet light, a full exploration of oxygen's history requires its own book (a good place to begin is

Nick Lane's *Oxygen: The Molecule That Made the World*). For a book focused on toxic chemicals, the debut of oxygen may seem beside the point: here we sit, breathing oxygen, without oxidizing from the inside out. And as with other naturally occurring chemicals and processes, we have yet to disrupt the earth's balance of atmospheric oxygen. But the evolutionary processes leading to oxygen detoxification provide an excellent example of life's resilience in the face of a dangerous pollutant: one that, in time, blanketed the earth. It is also a story that unfolds over billions of years and is played out through the evolution of single-celled life forms. As such, it is a triumph of unicellular life that benefits every cell in our complex, multicellular bodies.

Oxygen Basics

Like nearly all substances, there are beneficial and harmful amounts of oxygen. Oxygen turns butter rancid and raw meat green, it eats away at our iron pipes and steel bridges, and it ages us all. Yet in a life sustaining process, it also allows us to crack organic nutrients like sugars and fats into their smallest components—CO_2—and feast on their energy. Oxygen is also the *third* most abundant element in the universe, next to hydrogen and helium. By weight, oxygen accounts for roughly 66% of our body, 90% of water, and 21% percent of the earth's present atmosphere. And, as we learned in chapter 2, oxygen in the form of ozone absorbs UV light, allowing life to flourish on the planet's surface.

Life on Earth requires some amount of oxygen: if not for aerobic respiration (when oxygen is used for metabolism), then to create its most basic building blocks. All living things on Earth require carbon, hydrogen, oxygen, and nitrogen.[1] But the oxygen molecules in nucleic acids, proteins, fats, cell membranes, and other structural components are bound up with other elements, and unavailable as a chemical reactant. And those oxygen molecules that are bound together with two hydrogen atoms as water, or with myriad other elements (sometimes with the help of living things)[2] in a dazzling array of minerals, are also relatively stable and nontoxic. But there are also the reactive, radical forms of oxygen, which are some of the most destructive molecules on Earth. And it is the simple molecular oxygen, oxygen gas (or O_2), which under the right conditions gives rise to these toxic by-products. This is the oxygen that is of interest to us, and the form with which we are most familiar,[3] breathing in what amounts to several hundred liters of it per day.

During the oxidation of molecules containing carbon, energy is released. Whether the "burning" process is fast and direct, like the explosive combustion of tinder dry brush, or slower, indirect, and more controlled, as in the case of aerobic metabolism of carbon-based foods, the basic chemistry of oxidation is the same. We are able to use oxygen to metabolize food without self-destructing because our single-celled ancestors evolved a set of oxygen detoxification enzymes on which we rely today, and because, explains Nick Lane, molecular oxygen has an "odd reluctance" to react.[4] All chemical reactions, whether breaking apart or joining together, involve electrons: negatively charged subatomic particles. The joining of two molecules requires the formation of pairs of electrons, each with an opposite spin. Molecular oxygen, or O_2, however, has two unpaired electrons with parallel spin, resulting in a relatively unique molecule that has some difficulty interacting with other molecules (it cannot simply accept a pair of electrons, since one pair with opposite spin and one pair with parallel spin is an untenable situation).[5] But oxygen can interact with other molecules by first reacting in one of two ways. Circumstances (absorption of energy from heat or light, for example) can cause one of those electrons to flip, creating a highly reactive oxygen molecule called singlet oxygen: a reactive form in search of two electrons at once.[6] Any time a molecule requires an electron for stability, not to mention two, it has the potential for destructive behavior, attacking the electron bonds of a cell's organic molecules, including those of lipids and proteins. As such, singlet oxygen is one of the infamous reactive oxygen species, or ROS. Oxygen's alternate route to stability requires pairing those electrons one at a time: by combining with molecules capable of losing an electron or two without itself becoming unstable. In either case, the removal of electrons by oxygen is referred to as oxidation, and the molecule on the losing side is said to have been oxidized. Conversely, as oxygen gains electrons it is reduced, and the process is referred to as reduction: a complementary redox reaction. Metals like iron, manganese, and copper all readily interact with oxygen—a property that can be either fortuitous or disastrous for life evolving on a planet with increasing oxygen concentrations. So, if oxygen is both ubiquitous and destructive at the same time, at what point in life's history did it become relevant, and how did life evolve the means to defuse a pervasive toxicant?

Evolution of Oxygen

The history of oxygen on Earth is relevant for understanding life on Earth today and for considering life on other planets, all active areas of research. And while many details of oxygen's history on Earth are literally set in stone, the interpretations of these details are constantly being modified. Even with improvements in both biological and chemical analysis, much of oxygen's early history remains a controversial topic. As a biology major brainwashed to think that the study of rocks was only for jocks (never mind that I was also a female athlete), I missed a formal introduction to what I now realize is a fascinating and complex field. So I have no illusions of expertise in earth science. Fortunately, several good reviews and books are available for the interested reader.[7] Below I provide a limited review of some of the current research.

Considering the importance of oxygen to the evolution of life, its capacity to destroy life, and the role of life in oxygenating the planet, one cannot help thinking about the old chicken-and-egg puzzle. As so aptly put by the geobiologist Joseph Kirschvink and his student at the time, Robert Kopp, in reference to their own attempt to solve the puzzle, "O_2-evolving processes require O_2-mediating [detoxifying] enzymes to limit toxicity, while O_2-mediating enzymes are unlikely to evolve without a source of O_2."[8] At some point, whether very early in life's history or a billion years later, oxygen and life collided. At that point, oxygen would have presented a powerful directional selective pressure not only because of its toxicity, but also for its ability to drive the processes that crack apart the bonds of organic molecules, allowing the single-celled life forms inhabiting the earth to utilize every bit of pent-up energy through aerobic respiration. So which did come first, protection, production, or perhaps even benefit?

Despite the uncertainty surrounding the debut of O_2, there was no doubt plenty of oxygen on young Earth, only much of it was bound up in the planet's crust or flooding the planet's surface in the form of water. Should it surprise us then that water is the primary source of atmospheric oxygen? Perhaps. Water is a tenacious molecule, reluctant to release its oxygen, while oxygen itself can be reluctant to form the dioxygen, or O_2, on which we depend.[9] Breaking chemical bonds requires energy, and the interaction between water and the sun's light energy is an essential route to oxygen production. Whether this occurs directly in the atmosphere, at the water's surface, atop a glacier (abiotic options that we will discuss in a bit), or through biologically driven oxygenic photosynthesis, it is a remarkable event.[10] Water is the source of oxygen on our planet.

The evolution of oxygenic photosynthesis (whereby water and carbon dioxide are transformed into sugars and an oxygen by-product) required the formation of an extraordinary complex of chemicals now known as the "oxygen-evolving complex," or OEC: a set of four manganese ions and a calcium ion bridged together with oxygen.[11] This small collection of ions provided life with the capacity to split water. Alongside the evolution of DNA and RNA, it is arguably one of the most stunning developments in the history of the planet.[12] It literally changed the face of the earth, redirecting the course of evolution and greening the planet with a range of photosynthetic life, from the smallest bacteria to the tallest redwoods.

Thanks in large part to the earliest successful photosynthesizers, the earth's oxygen concentrations rose from trace amounts to less than a few percent of current concentrations roughly 2.4 billion years ago, during the Great Oxidation Event (or GOE);[13] and rising again, over a billion years later, to current levels. Woe to life unprepared—if there were such creatures. Back in the 1970s, the visionary scientist James Lovelock suggested that this clash between life and oxygen was grounds for catastrophe, writing, "When oxygen leaked into the air two aeons ago, the biosphere was like the crew of a stricken submarine, needing all hands to rebuild the systems damaged or destroyed and at the same time threatened by an increasing concentration of poisonous gas in the air. . . . The first appearance of oxygen in the air heralded an almost fatal catastrophe for early life."[14] Yet instead of a cataclysmic meeting of oxygen and life, what if the introduction occurred more gradually? What if life had the opportunity to prepare? Writing about this alternative scenario, Lane suggests that "the oxygenation of the earth seems to have proceeded in a series of sharp jerks or pulses," such that after the initiation of the GOE, nearly a billion years followed before oxygen levels began to rise significantly.[15] And even before the GOE, small amounts of oxygen species may have been available to life.

The fits and starts of the earth's oxygenation around the time of the GOE are observed in geological formations, including banded iron formations (BIF). Recall that oxygen rusts iron, and the primordial oceans were rich in the most soluble form of iron, ferrous iron (or Fe^{2+}). Yet upon encountering oxygen, ferrous iron readily loses an electron, transformed through oxidization to ferric iron (Fe_{3+}) or rust. As oxygen availability increased, a mix of insoluble compounds, including rust, began settling out of solution as iron-rich deposits or bands, which now, in addition to preserving oxygen's history, serve as a source of iron ore. (There are also pre-GOE banded iron formations, and the reasons for BIF formation at a few post-GOE times are not well understood.)[16] As iron "sopped" up the toxic gas, seawater gradually converted from an iron-rich environment to an iron-poor environment. This change likely presented our ancestors with yet another hurdle, as iron had by then become incorporated into their single-celled bodies as an essential metal. (Iron and other essential metals will be discussed in greater detail in chapter 4.) Recent analysis of chromium isotopes found in banded iron formations also raises interesting early Earth scenarios. One interpretation suggests that oxygen concentrations may have dropped to nearly trace levels some half a billion years after the initiation of the GOE, while another suggests that small amounts of oxygen may have been available, possibly from photosynthesis, at least three hundred million years prior to the GOE.[17] If early oxygen levels fluctuated rather than rising suddenly, there may have been a "grace period," during which oxygen served as a gentle selective pressure without annihilating whole populations—including those first photosynthetic cells pumping out oxygen in excess. Before oxygen powered the microbial masses during and after the GOE, did it mingle with small populations of cells? If so, where did it come from, and could there have been enough of this new environmental stressor to prime life for an aerobic existence?

Considering the abiotic production of oxygen by ultraviolet radiation, we return to Joseph Kirschvink, who coined the term "Snowball Earth," and who suggests that large enough sources of oxygen may have become locked up during major glaciations. According to Kirschvink, the planet plunged into a deep freeze just before the GOE. Now, water is water no matter the phase, whether vapor, liquid, or frozen into glaciers. When UV radiation reacts with water, one outcome is the production of H_2O_2, or hydrogen peroxide, a reactive oxygen species.

Because hydrogen peroxide has a freezing point very near that of water, suggests Kirschvink, it is possible that glacial compression allowed hydrogen peroxide to concentrate within the glacier. When hydrogen peroxide meets reactive metal salts, it decomposes into oxygen and hydrogen. As glaciers melted, frozen, concentrated H_2O_2 would have had some opportunity to interact with metal salts, eventually releasing oxygen into the surrounding water, providing, perhaps, just enough oxygen to exert selective pressure on local populations.[18] (Similarly, exposures to small amounts of certain contaminants like metals, antibiotics, or even dioxins today may select for life resistant to higher concentrations, particularly if existing proteins or enzymes are available to be co-opted or ramped up, defusing the situation at hand and allowing life to proceed.) Nick Lane goes even further, suggesting a role for ultraviolet light–induced H_2O_2 production well before Snowball Earth.[19] In either case, abiotic (non-photosynthetic) sources of small amounts of O_2 may have influenced life's early environment, providing at least some species with a taste of things to come, and perhaps the opportunity to evolve protections well before the GOE.[20]

Synergy

As discussed in chapter 2, UVB is fully capable of damaging DNA on its own. It is, in a sense, a "complete" toxicant, as its mechanism of action is direct and independent of metabolism or the presence of other chemicals. On the other hand, oxygen species like H_2O_2 require some conversion before becoming highly toxic. Moderately toxic on its own, H_2O_2 becomes radically more destructive in the presence of some forms of iron. This combo provides us with a very old example of toxic synergy in a chemical mixture, where the toxicity of the combination is far greater than that of the individual chemicals.[21]

As we know, before oxygen's planetary invasion, the oceans were rich in ferrous iron. Because ferrous iron is not completely oxidized, it will donate an electron to H_2O_2, producing one of the most reactive and destructive ROS known, the hydroxyl radical ($^{\bullet}OH$). This reaction, discovered by the chemist Henry John Horstman Fenton in the late nineteenth century, turns a relatively benign solution of hydrogen peroxide into a ferocious oxidizing agent.[22] It is a process used today to break down organic matter in wastewater treatment—and a reaction that has been the bane of life for eons.

Now let us consider the fate of some ancient, single-celled organisms living nearby a melting glacier or a small pool under the sun's full bore on a very young Earth. Some of the earliest and simplest defenses against abiotic-derived ROS, like those for UV light, were likely physical.[23] Protection by burrowing into anoxic mud, remaining deep undersea, or moving away from high concentrations of H_2O_2 may have prevailed at first. Yet with a good deal of time, and inadvertently taking advantage of random mutations, life eventually evolved detoxification systems that allowed it to slowly live in the open air. Perhaps life benefited from improved sensors that enhanced avoidance, or sacrificial extracellular membranes that took a toxic hit while protecting the inner cell, or metabolic pathways that ended with the deposition of the toxicant into oxygen-rich molecules like collagen, or protective molecules like the hard calcium carbonate shells excreted from many invertebrates today.[24] One thing we do know is that eventually enzymes capable of squelching oxygen's toxicity emerged. This is arguably the single most important evolutionary change that allowed life to coexist in a world polluted with oxygen.

When the Fenton reaction happens outside a cell, chances are outer membrane molecules might collect the toxic product. But hydrogen peroxide is one of the more soluble oxygen species, capable of diffusing across cell membranes. Should H_2O_2 have entered the cell and encountered a molecule of ferrous iron, the consequences would have been catastrophic, with wholesale destruction of essential proteins, lipids, and nucleotides. In this case, external protection would not be enough. And this conundrum brings us to the evolution of catalase, the enzyme that destroys hydrogen peroxide.

Catalase, a Saving Grace?

Today, while the great majority of species living on the earth's surface thrive in an atmosphere of 21% oxygen, many, including humans, maintain much lower concentrations of oxygen at the cellular level. In fact, tissue oxygen pressures for many animals, including mammals, tend to resemble the low-oxygen conditions encountered by some of the first eukaryotes rather than the present atmospheric conditions.[25] This may suggest that the suitable oxygen balance for life has changed very little over time. Yet not all living things made their peace with oxygen. Many species of microorganisms, from *Clostridium botulinum*, which haunt our canned goods, to deep-sea bacteria colonizing hydrothermal vents, along with a few recently discovered deep-sea-dwelling animals, flourish in the absence of oxygen, and perish in its presence.[26] Could the same be said for one of our more intriguing forebears, the grand dame herself, LUCA, our last universal common ancestor?[27]

Was she or wasn't she an anaerobe, and how would we know? Given the timeline of the GOE and the general acceptance that early life was anaerobic, it had been assumed that LUCA was an anaerobe. As we know from chapter 2, however, the recent genomic analyses by Christos Ouzounis and colleagues provide insight about LUCA's complement of proteins and enzymes—and possibly her environment. Those data reveal that LUCA, a pre-GOE life form, may have produced at least two enzymes, catalase and superoxide dismutase, or SOD. These enzymes would have provided some capacity to detoxify ROS like hydrogen peroxide, and in the case of SOD, superoxide (O_2-), a reactive product of oxygen respiration.[28] Regardless of whether oxygen detoxification was the original function, both enzymes are now considered essential for detoxification of reactive oxygen species for nearly all oxygen-respiring life, including us. Perhaps, if hypotheses about early exposures to H_2O_2 are correct, LUCA's capacity to detoxify the chemical is not surprising. But what to make of an enzyme used to defuse a product of respiration? Pondering LUCA's capacity to produce SOD, Ouzounis and colleagues speculate that LUCA may not only have experienced oxygen exposure, but may have even dabbled in aerobic respiration.[29]

In the end, we may never know whether LUCA experienced oxygen. While some analysis suggests yes, other work claims no.[30] At the very least, we know that some time after LUCA, there was plenty of oxygen in the air and in the water. And it is likely that life was prepared with catalase, perhaps one of the oldest enzymatic defenses to reactive oxygen, and an enzyme that remains essential for oxygen detoxification today.

In Defense of Oxygen

Returning from a quick jog around the neighborhood, I consider the excessive amounts of oxygen I've inspired over the past forty minutes and think about its brief and potentially destructive course through my body. After diffusing across the thin cellular membranes into the myriad capillaries that perfuse my alveolar sacs, it will join with hemoglobin—a combination of iron bound with protein, and another vestige of life's early struggle to survive oxygen toxicity. This metallo-protein (a metal combined with a protein) travels the body, to each and every cell, exchanging oxygen with carbon dioxide. Once there, the cell's mitochondria (often referred to as "powerhouses of the cell") engage in aerobic respiration as they oxidize glucose, releasing large amounts of energy in the form of ATP, the cell's energy currency. As this happens, hazardous by-products in the form of ROS will be released. Some ROS will damage my cells, effectively aging me[31] and causing me to ponder the wisdom of this thrice-weekly torture. Yet I can take some solace that I am fairly well equipped to battle oxygen toxicity. Within each cell, catalase (which incidentally is plentiful in my oxygen-transporting blood cells), SOD, and other antioxidant enzymes will tame some of those ROS, transposing destructive oxygen radicals back into water molecules from whence they came.

Catalase is one of the most efficient enzymes on Earth. Although it may seem counterintuitive considering iron's role in activating H_2O_2, catalase contains iron. Yet here are some startling figures that provide perspective: in the absence of iron and iron-containing proteins, the conversion of hydrogen peroxide to water and oxygen would take weeks; should it encounter iron simply bound to a protein (oxygen-transporting heme, for example), the reaction occurs one thousand times faster; and if it encounters iron in the form of catalase (with its iron-containing heme group), the reaction time increases such that in just one second, one catalase molecule will completely break down millions of hydrogen peroxide molecules, thereby preventing the production of millions of hydroxyl radicals.[32]

The changing role of iron presents us with a fascinating glimpse of evolutionary processes that co-opted a rogue hydroxyl radical generator, bound it up with a protein, and then perhaps surreptitiously employed the metal-protein combination (metalloprotein) in the controlled degradation of hydrogen peroxide. Iron is now so important for life, as mentioned earlier, that it has become an essential nutrient (and sequestering iron from invading microbes is one form of the body's defense).[33] It is not known whether catalase evolved from a metalloprotein serving another purpose. What we do know is that a highly integrated and finely tuned system of antioxidant enzymes has evolved to the extent that aerobic life now thrives in what is a relatively stable concentration of atmospheric oxygen (depending on altitude). In addition to catalase, we also rely on other "frontline" enzymes, including peroxidases, which degrade H_2O_2 and SODs. As with many enzymatic reactions, sometimes these enzymes depend on each other to see a reaction through to completion. Protecting aerobes from respiration requires both the SODs, which "dismutate" the superoxide generated during respiration into H_2O_2, and the "closers," catalase and peroxidase, which complete the detoxification process.

As we will see throughout this book, rather than relying on one remedy, life is full of backups and redundancies. Besides enzymes and physical protections, the struggle to both utilize and tame oxygen includes a host of adaptations and inventions. The evolution of a symbiotic relationship leading to mitochondria, for example, may have served to sequester respiration in these well-equipped organelles. The packaging and use of H_2O_2 in our immune system both sequesters ROS and directs its release as a form of protection from invading organisms. The internal production of molecules like glutathione, one of the most important internally produced antioxidant molecules, donates electrons to reactive oxygen, thereby quenching that volatile element and protecting nearby tissues and molecules. Yet despite all the internal efforts toward protection, many species also require the ingestion of a host of dietary antioxidants. From lycopenes in tomato-based products to resvesterol in red grapes and epicatechins in dark chocolates, dietary antioxidants abound in plant-derived foods. These additions likely became increasingly important as plant-animal relationships evolved.

The topic of antioxidants is close to the hearts of many people invested in "healthy living" today, generating books and articles ad nauseam. A quick trip to Amazon.com reveals more than fifteen thousand books listed under antioxidants. Rather than getting lost in the antioxidant literature, let's instead return to the simpler concept of linking a system's evolutionary history to the modern-day environmental pressures testing that system.

Marathon Man

Oxidative stress is a common term for damage caused by an imbalance between reactive oxygen and the capacity of a cell (or body) for detoxification, whether the result of normal respiration, poor diet, exposure to certain artificial pesticides, or genetic history. Many pesticides, including organochlorines and organophosphates, cause oxidative stress, either directly through inhibition of antioxidant enzymes or indirectly through depletion of other antioxidant molecules, resulting in concentrations of ROS well beyond those for which some individuals or populations are prepared.[34] Air pollutants, including metals and the very smallest particulate materials ($PM_{2.5}$, for example), provide ample opportunity for ROS generation in our lungs, heart, and other tissues.[35] Scientists are just now considering the potential risks associated with ROS generation by nanoscale particles, a rapidly expanding industrial sector. Like antioxidants, each one of these topics could fill a book. On a more manageable level and more to the point, there is (at least) one example of a species being exposed to amounts of oxygen well beyond those commonly encountered in its evolutionary history. And, of course, that species is us, *Homo sapiens*. From exposing premature infants to oxygen concentrations double or triple atmospheric concentrations, a common practice in the past, to our relatively current fitness craze, humans are pushing their antioxidant systems to the limit, challenging their bodies to an oxygenic evolutionary mismatch.[36]

Referring to the aerobic exercise craze that took hold in the United States decades ago, the exercise physiologist Robert Jenkins commented, "There probably has never been a time in human history when such a broad cross-section of humans have consumed such large doses of oxygen for concentrated periods of time. Primitive hunters stalked game; they probably did not spend a half-hour or an hour each day running at a pace equal to 70–80% of their maximal oxygen intake."[37] So let us reconsider the excess oxygen inspired while jogging, cycling, or engaging in prolonged periods of just about any aerobic exercise. Here are just a few sobering figures borrowed from a recent publication on oxygen stress in elite female athletes: while exercising, a body's oxygen consumption may increase upward of twenty-times compared with resting; oxygen in certain muscles may increase one hundred times over resting concentrations; and while the great majority of oxygen will help turn glucose into energy, two 2%–5% will end up as ROS, a consequence of electron leakage from respiration's electron transport chain.[38]

Our ancestors who farmed, hunted, kneaded bread, chopped wood, and walked miles to their neighbors' houses did not sit behind glowing screens of one kind or another for eight hours a day. And, noted Jenkins, they didn't spend lunchtime on the treadmill. Our own evolutionary history was likely shaped by moderate or perhaps sporadically active populations—and the tendency toward moderate physical activity is likely reflected in our genes.[39] We have not (yet) evolved to be sedentary beings. Nor are we adapted to quickly flipping between sedentary and highly active lifestyles, including weekend jogging or high-intensity community soccer games. So how does our finely tuned antioxidant network respond to this sporadic but increased oxygen intake?

Not surprisingly, there is an abundance of evidence that exercise increases ROS formation. The consequences can be measured in damage to lipids, proteins, and DNA.[40] Yet we know that well-trained athletes, who continually test the balance between ROS and antioxidants, tend not to suffer from massive oxidation or literal burnout. If we consider the deeper evolutionary history, many species sometimes require protection from sudden or ramped-up activity: racing to escape from or capture predators; prolonged migrations after periods of relative inactivity; involuntary exposures to rapid changes in ambient oxygen in less-mobile organisms. It might not be surprising then, that antioxidant systems are flexible in their responsiveness to ROS. That is, following exposure to nonlethal amounts of ROS, antioxidant enzymes, including SOD and catalase, are not just mobilized but also may be up-regulated, resulting in increased gene expression and protein production.[41] In other words, not only does a moderate amount of exercise prime and protect the body, preparing it for more ROS (some have even claimed that "moderate exercise is an antioxidant"), but ROS also acts as the molecular signal turning on and off appropriate genes.[42] This adaptive response to low doses of a potentially damaging molecule—a phenomenon for which it seems the more we look, the more we find—is referred to as hormesis (the role of hormesis is discussed in greater detail in chapter 8). This system is so finely tuned that some researchers and elite athletes now question the efficacy of antioxidant supplements, suggesting that rather than protecting against exercise-induced ROS, they may instead interfere with the natural mobilization of antioxidants.[43]

While we have modulated our exercise habits well beyond anything our immediate ancestors could have fathomed, we are not the only species to alternate between periods of inactivity and intense and sustained aerobic exercise. Wildlife, particularly migratory species like birds, do it twice a year, while salmon may do it only once in a lifetime. How do they cope? Like us, hypothesizes the biologist David Costantini, birds might suffer oxidative stress at the outset of their migration, but by the time they've "trained" they may be better able to withstand any ROS assault.[44] Costantini and other biologists and ecologists are rapidly expanding the nascent field of antioxidant or oxidative stress ecology as they consider the role of antioxidants and oxidative stress in everything from sexual selection to senescence in wildlife.[45] Additionally, suggests Costantini, those who do migrate may have evolved more robust antioxidant networks. But producing proteins or large molecules that are not directly related to life's ultimate goal of reproduction could present a trade-off in terms of energy allocation—a fact not lost on ecologists like Neil Metcalfe and Carlos Alonso-Alvarez. In fact, they suggest that because some antioxidant activities must be induced, or turned on, battling these ROS is most likely a costly process.[46] As we shall see throughout this book, many other defensive mechanisms, depending on their raison d'être, require the turning on and off of myriad genes—for that reason, as perhaps with all battles, defensive preparedness can be expensive.

Finally, before we jump to conclusions about broadening the purview of protective up-regulation by ROS generators, let us briefly consider one example that is not necessarily part of our evolutionary history. Recall those very small particulate air pollutants, the PM_{10} and smaller. While not necessarily new to life, they are now present in some regions, particularly in congested cities, in concentrations well above what could be considered "background" or "natural." Once inhaled, PM_{10} not only enter the bloodstream, but as they travel throughout the body from lungs to the heart, they may also leave a trail of ROS. Particulates like PM_{10} create ROS by virtue of their relatively large reactive surface areas and, in part, because some contain reactive metals, including iron. A recent study, exposing whole cells to nonlethal concentrations of PM_{10}, revealed the complicated nature of response, as ROS increased in these cells, while catalase and SOD activity was reduced.[47] Yet cell survival was not affected. If the cells didn't die, does that mean a little bit of PM_{10} is harmless or, should those antioxidant levels bounce back with a vengeance, possibly even beneficial in the long run? Not necessarily. When subsequently challenged with less than lethal doses of H_2O_2, which can occur simply through increased activity or because of a range of respiratory diseases, cells that survived the initial assault died. While we ought not interpret these results too broadly, it would seem that, in this case, rather than priming cells for a response, the combined exposure to PM_{10} and other ROS set them up for a lethal condition. In a larger context, such combined exposures even to sublethal concentrations of like-acting pollutants (in this case, ROS generators) may be sufficient to overwhelm defenses, creating conditions ripe for an evolutionary mismatch. This topic begs for further research, and will be touched on as we consider the impacts of contemporary pollutants in chapter 10.

Oxygen, like sunlight, is both a necessity and a life-threatening toxicant. Yet billions of years of evolution under the influence of oxygen has resulted in a responsive system that is for the most part capable of protecting life from, and also capitalizing on, a toxic chemical. In the case of ROS, we now know that tens of thousands of genes may respond—and scientists are just beginning to explore the broad reach of ROS and oxidative stress.[48] Over the past decade or two, as researchers turned their attention to the role of ROS in aging, reproduction, and disease, they are finding that ROS are not simply toxic by-products of oxygen respiration, but also essential and important signaling molecules, adding yet another layer to life's paradoxical relationship with oxygen. This double-edged sword is a common theme in toxicology—and one that will be explored further as we next consider the role of metals on the evolutionary history of life's response to chemical hazards.

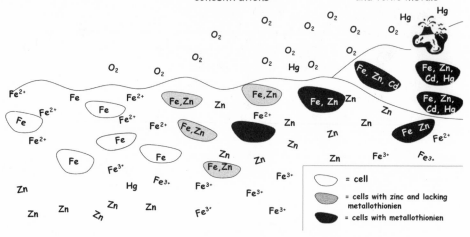

Then

Low or no oxygen
meant highly available
soluble iron (Fe^{2+})

Toxic Evolution

Changes in oxygen made zinc
more available and iron scarce;
metallothionein (MT) may have
helped maintain zinc
concentrations

Now

Zinc and other metals are
now essential and MT
binds with both essential
and toxic metals

= cell

= cells with zinc and lacking
metallothionien

= cells with metallothionien

Metal availability changed over the course of life's evolution.

Chapter 4

Metal Planet

Earth formed as a heavy metal planet.

Robert Williams and J. J. R. Frausto da Silva

Life's "addiction" to iron is thought to reflect this early evolution in an iron-rich reducing environment.

Jennifer S. Cavet, Gilles P. M. Borrelly, and Nigel J. Robinson

Just as we are reminded of our aquatic origins when we taste the salt of our tears, each time we reflexively lift a cut finger to our mouths, the metallic taste of blood is a reminder of our origins in an iron-rich environment. We carry many other metals, most notably copper and zinc, which also refer us back to elements available to our distant ancestors. Metals are fascinating because while some are simply toxic like lead, others like zinc and copper are both essential *and* toxic. For humans, metals are essential not only physiologically, particularly for proper protein function, but also economically, as these elements played key roles in the advancement of civilizations. Our idolatry of gold, silver, titanium, and other metals has, for better or worse, dramatically changed the types and amounts of metals available in the biosphere. Yet, when it comes to the history of the chemistry of metals on Earth, humans are far from the most important agents of change. We know that life evolved in an iron-rich aquatic environment relatively devoid of other metals and minerals now recognized as essential. We also know that as the earth's atmosphere changed with the introduction of oxygen, dramatic changes in metal availability occurred as well. Metals once plentiful turned insoluble, dropping out of solution and away from life's grasp, while others became increasingly soluble and washed into the sea, more available than ever before.

One of the most notable changes was the oxidation of iron from soluble and bioavailable ferrous iron (Fe^{2+}) to insoluble and less available ferric iron (Fe^{3+}, or rust), altering life's ability to access what had become an essential metal. Conversely, the earth's oxidation caused metals like zinc and copper to trickle from rocks and minerals into nearby coastal regions, presenting new opportunity for life, while at the same time contaminating its environment with potentially toxic metals. Changes in metal availability created new selection pressures for living organisms. They responded by finely tuning processes that balanced the need for essential metals while protecting against toxicity.

The majority of naturally occurring elements are classified as metals. From soft to hard, abundant to rare, essential to highly toxic, metals include elements from calcium to sodium and from copper to zinc.[1] One shared characteristic of metals is their propensity to lose electrons, becoming positive ions (or cations). (Oxidation state plays an important role in determining the toxicity of many metals.) As such, they are willing participants in reduction/oxidation (redox) electron transfer reactions. As discussed in earlier chapters, electron transfer is an essential process in the production of ATP, the so-called currency of life. Therefore, metallic elements are often glinting at the heart of life-sustaining reactions, including photosynthesis and respiration. Their ability to easily lose and then gain electrons, combined with their malleability and strength, provides us with not only cast iron skillets and kitchen knives but also the route through which electrons flow from a coffee shop outlet to my laptop. The human quest for economically important metals such as gold, aluminum, and platinum—in addition to metals now popular in the hi-tech industry like neodymium and tantalum, and the inadvertent releases of metals like mercury from coal—has caused a modern redistribution of metals around the globe. This has also created unprecedented environmental challenges. And so the biochemical mechanisms that may once have kept essential metals in check, or locked up the occasional toxicant, are now frontline defense mechanisms for a diversity of species. As with oxygen and UV light, the combined natural history of metals and the evolutionary history of their basic utility as redox active elements provide us with another example of life's capacity to create, deal with, and even take advantage of environmental change. Understanding this interplay between environment, metals, and evolution provides insights overlooked when we focus solely on life's present condition. And so we begin with an overview of metals throughout the early history of Earth (and life).

The Metals Exchange

Stumbling across the granite peaks of New Hampshire's White Mountains on a cool, clear October day, I consider the relative stability of the weathered domes, eroded boulders, and rock slides that characterize one of the oldest mountain chains in the world. Granites can contain dozens, perhaps hundreds, of different minerals (metals in combination with other elements), the majority of which are quartz, feldspar, and mica, along with, depending on location, much smaller amounts of elements such as cobalt, yttrium, vanadium, nickel, neodymium, tantalum, and uranium. The mountains of New Hampshire have bared silent witness to more than one hundred million years of change. Lichens, trees, wildlife, and even humans come and go, but the Granite State is here to stay (at least for a while). Yet while we may think of these chemical components as relatively stable in contrast to the ever-changing business of life, the availability of both the types and amounts of minerals, and metals in particular, changed dramatically with the advent of photosynthetic life. Proposing a "new framework" for mineral evolution, the geologist Robert Hazen suggests that although only a dozen or so minerals existed at the time of the solar system's formation, and some 250 minerals typically characterize rocky planets devoid of life in the present day (as Earth once was), life is responsible in large part for the great diversity of minerals observed on our planet today.[2] In an article in *Scientific American*, Hazen writes that of the more than 4,400 known minerals, roughly two-thirds can be attributed to the production of molecular oxygen by living organisms during the Great Oxidation Event (or GOE)—a change that allowed oxygen, formerly limited to one small set of mineral forms, to take part in multitudes of other combinations.[3] Oxygen-producing life permanently altered mineral and metal availability, contaminating their own environment not only with oxygen but also with newly available metals.

In their book *The Chemistry of Evolution: The Development of Our Ecosystem*, Robert Williams and João Frausto da Silva note that Earth is a heavy metal planet, and has been ever since its earliest high-temperature incarnation, when molten iron collected at its core. One benefit of the earth's iron core and rotational motion, as discussed in chapter 2, is the magnetic field, which shields the earth from the sun's intense radiation. While the planet's iron core protects life from external radiation harm, its surface—the atmosphere, aquatic environments, and crust—is where earth meets life. When atmospheric oxygen caused metals and minerals to seep from the crust to the oceans, life had to adapt. The relevance of mineral availability in life's evolution can be illustrated by considering the Achaean sea—rich in potassium, magnesium, calcium, manganese, and iron, all of which are essential for life today. Meanwhile, prior to the GOE, metals now integral to almost all forms of life would have been available at one-thousandth of today's concentrations, and likely were not essential—at least for our Achaean ancestors. About these changes, Williams and Frausto da Silva write, "The oxidation in sequence of the elements had a profound influence on the evolution of life, but is far from complete."[4] Using the trail of elements left behind by the earth's oxidation, the two elucidate the role of elemental change both in the environment and in life (discussed in detail in *Chemical Evolution*). One important global change was the conversion of Earth from a "sulfur world" to an "oxygen world." "Sulfur," writes Williams and Frausto da Silva, "played a major role in the beginning of life and then subsequently metal sulfides and their subsequent reactions have had a major influence on evolution."[5]

On a young Earth, sulfur, particularly hydrogen sulfide (H_2S) and sulfur dioxide (SO_2), was (and still is) associated with black smokers or deep-sea vents and with volcanic eruptions. While some sulfides are quite soluble (sodium, calcium, magnesium, and iron sulfides, for example—all common elements in the early Achaean sea), others, particularly sulfides of cobalt, cadmium, nickel, copper, and zinc (depending on local conditions), were insoluble and relatively unavailable. As oxygen converted insoluble sulfides to soluble sulfates, those once insoluble metal ions eventually washed into the sea, and at some point made their way into life. As the availability of metals, including Co, Ni, Zn, Cu, and Cd, increased, soluble iron sulfides turned into insoluble iron oxides.[6] Sulfur chemistry clearly affected the evolution of life through its association with minerals and metals in the external environment, but what of its role in life? Although not one of the five elements required for DNA and RNA (C, H, O, N, and P), sulfur is certainly next in terms of importance. It is critical not only for its role in maintaining the homeostasis of essential metals, but also for its ability to combine with several species of metals.[7]

So how did life survive dramatic changes in metal availability? It evolved finely tuned systems to scavenge, pump, bind, transport, and sequester metals that became essential for life processes, while developing detoxification schemes for those metals either too reactive or too nonreactive to be useful. Those metals become toxic, taking the place of functional metal sites in an irreversible manner and rendering dysfunctional the proteins in which they were embedded. For many metals, toxic and essential are two points along a single continuum. Though we may never know which end best describes the initial interactions between any given metal and life, both are clearly important, and so before focusing on the toxic properties, we shall consider the elements in their essential roles.

Bare Necessities: Essential Minerals and Elements

Despite its changing environment, or perhaps because of it, life invested in a set of metals that have been retained through billions of years, including vanadium, molybdenum, cobalt, copper, chromium, iron, manganese, nickel, and zinc—and which over time became essential.[8] A metal is deemed essential when concentrations below the range of "adequate intake" result in impaired function.[9] Often these are metals straddling the border between toxicant and nutrient, and while some metals are required in relatively high concentrations, trace amounts of others are sufficient. For example, human bodies have on average about four grams of iron and two grams of zinc, while concentrations of cobalt and nickel are about one thousand times lower.[10] Additionally, while most species need iron, the requirements for other metals have been gained and lost with the diversification of life and the environments it inhabited. Zinc, for example, is required in relatively larger amounts in eukaryotes (which evolved primarily in an oxygenic and soluble-zinc world) in comparison with prokaryotes.[11]

Understanding why any one particular metal became essential requires some consideration of its availability. This holds true even for metals needed (and available) in the smallest amounts. Yet estimating availability requires relatively detailed knowledge of the local environment in which life evolved: knowledge that is only as good as the current analysis. Molybdenum is a case in point. This rare metal played a small role in the short-lived hypothesis of directed panspermia: the deliberate infection of Earth by intelligent beings from another planet.[12] This seemingly zany idea was once championed by the biochemist Leslie Orgel and the Nobel Prize winner Francis Crick, two well-respected scientists who contributed significantly to our current understanding of the genetic basis of life. While acknowledging their argument as "weak," they cited life's requirement for molybdenum as evidence of Earth's infection. In their 1973 paper on "Directed Panspermia," the authors wrote:

> The chemical composition of living organisms must reflect to some extent the composition of the environment in which they evolved. Thus the presence in living organisms of elements that are extremely rare on the Earth might indicate that life is extraterrestrial in origin. Molybdenum is an essential trace element that plays an important role in many enzymatic reactions, while chromium and nickel are relatively unimportant in biochemistry. The abundance of chromium, nickel, and molybdenum on the Earth are 0.20, 3.16, and 0.02%, respectively. . . . If it could be shown that the elements represented in terrestrial living organisms correlate closely with those that are abundant in some class of star—molybdenum stars, for example—we might look more sympathetically at "infective" theories.[13]

Explaining molybdenum's relative scarcity in light of its role as an essential metal might have required (or supported) some creative historical thinking. But it is now thought that there was more molybdenum than nickel or chromium in the oceans, where many scientists currently believe early life developed.[14] Even so, just because a metal is available does not mean it will become essential for life. Otherwise we might all require a bit of mercury, cadmium, or titanium—all available, albeit in small concentrations. (One of these elements may someday be found to be essential in exceedingly small amounts for one or more organisms.)

Conversely, we might consider iron, an element that was once readily available in the earth's crust, but now exists in relatively inaccessible forms. Two and a half billion years of evolution in an environment enriched with the metal, perhaps combined with its intrinsic chemical properties, have made it irreplaceable. Iron became so entwined with life that replacing it with another, more abundant metal appears to have been untenable for most species, even as iron became rare.[15] While we acquire iron from our diet, some microorganisms have evolved mechanisms to cope with iron-deficient environments, even if only temporarily. Cyanobacteria, for example, produce iron-scavenging structures called siderophores, while other microbes and some algae faced with low (or even no) iron replace proteins requiring iron with flavodoxins, a less efficient but workable substitute that allows survival until iron conditions improve.[16] More directly related to human health, the limited amount of iron available in our own cells may incite competition between host and pathogen. Some pathogenic microorganisms that prey on humans rely on siderophores, allowing them to scavenge iron even as our cells sequester the precious metal away. It is not hard to envision how capitalizing on this relationship might lead to new antibiotics or treatments.[17]

Zinc is another metal of critical importance for most species. Essentially unavailable in the open ocean at the dawn of life (although abundant following the GOE), zinc now provides stability and structure to proteins involved in cell replication, repair, and metabolism—all surely required by our Achaean ancestors. So given its scarcity in the Achaean ocean, how can we explain life's dependence on zinc? Like molybdenum, this conundrum has prompted alternative scenarios for life's origin. Rather than turning to distant planets, the biophysicists Armen Mulkidjanian and Michael Galperin point to Earth's hydrothermal vents as the most likely incubator for life.[18] Others have proposed this location based on additional components thought necessary for the inception of life.[19] In two papers titled "On the Origin of Life in the Zinc World," Mulkidjanian and Galperin lay out their hypothesis that life may have originated near zinc-sulfide precipitates associated with hydrothermal vents. Zinc would have become solubilized and biologically available through its role in catalyzing abiotic photosynthesis.[20] Challenging this explanation of life's zinc addiction are Christopher Dupont and Gustavo Caetano-Anolles. Their genomic analysis of metal-binding proteins confirms that "the oldest metalloproteins were almost certainly Zn binding and are ubiquitous in extant life,"[21] a conclusion that curiously seems to support the hydrothermal zinc hypothesis. But they also point out that zinc-binding proteins were likely rare, and that some species today (as discussed above, with iron) manage to survive when zinc concentrations are low. Even as zinc infiltrated seawater, and before oxygen concentrations stabilized zinc, concentrations likely fluctuated. Given this fluctuation, it might be argued that zinc was central in the formation of early life. Proteins requiring zinc may have had some capacity for flexibility, including being able to use other elements, like iron, and returning to zinc when conditions settled.[22] Proteins able to swap one metal for another and remain functional are referred to as cambialistic. Considering the physical chemistry of some metals, their ability to stand in for one another is not surprising.

Flexibility is obviously beneficial when environmental conditions render useful metals scarce.[23] Some superoxide-dismutase enzymes, for example, are able to exchange iron for manganese under certain conditions. But one can also imagine that the presence of a *less efficient* yet chemically similar metal, if it became abundant, would lead to disastrous consequences if uncontrolled substitution were to take place. Consider the essential metals—Mn, Ca, Fe, Co, Ni, Cu, and Zn—and the similar yet toxic metals—Cd, Hg, Ag, and Pb. Should any of the latter group become more prevalent, they might gain entrance into exposed cells, using pumps and other mechanisms evolved to regulate and transport their "look-alikes." Once inside, these metals may then interact with proteins, nucleic acids, and other biomolecules, replacing essential elements and disrupting normal function.[24] Even among the essential metals, the replacement of one metal for another, for instance Cu^{2+} for Zn^{2+}, can wreak havoc.[25] For these reasons, it is critical to maintain the balance or homeostasis of essential metals, and also to develop protections against "nonessential" elements. Metals must be tightly controlled, and they are.[26] Copper-binding proteins, for example, ensure that there are essentially no free copper ions inside cells (because they can be so toxic).[27] Metals have become so central to cellular function that the collection of metal-binding proteins (referred to as the metallome) accounts for over 30% of all proteins in the cell. Metals are known to be involved in over 40% of enzymatic reactions, and metal-binding proteins carry out at least one step in almost all biological pathways.[28] Metals, particularly zinc, copper, and iron, may help proteins assume their active, three-dimensional conformation. One common location for metals associated with proteins, including copper, zinc, and iron, is within protein folds. The evolutionary history of folds associated with manganese and iron suggest they *may* have appeared before those associated with zinc and copper—in accordance with their bioavailability prior to the GOE.[29]

As the earth's environment changed, so too did the metallome. It is not hard to imagine that once life moved from sea to land, the challenge became maintaining homeostasis, as elements that were diffusely available in seawater were only sparsely or sporadically available on land. Living things not only had to be protected from metals that might occur in higher concentrations in their new homes, but also needed to scavenge for metals that had become rare yet were essential to their cellular function. To do the job, life continued to evolve a system of cellular pumps that controlled movement of metals both into and out of the cell, and chaperones to control availability and concentration in the cell. One prominent family of proteins involved in maintaining balance are the metallothioneins (MT): a family of metal-binding proteins first identified in association with cadmium, now known to exist in a diversity of species, from microorganisms to humans, and with a multitude of roles, including binding metals. When toxicologists think of metal contamination, we often think of metallothionein.

Balance

A little over twenty years ago, as a myopic environmental toxicologist focused on the virtues of cytochrome P450s (or CYPs, as they're called now), I paid little attention to their drab sister of detoxification, the metallothioneins. Back then it seemed these relatively simple metal-binding proteins were the purview of invertebrate toxicologists, who were most at home poking around in cadmium-contaminated coastal waters or digging for earthworms around old foundries. The CYPs, in contrast, *did* something. They metabolized polyaromatic hydrocarbons, hastening their excretion. They deactivated (and in some cases activated) drugs and chemicals. More important (at least to those of us making a living off CYPs), they could be used to indicate exposure to contaminants like PCBs and dioxins, creating highly fundable study areas. How could a comparatively passive protein that bound up metals be of any interest? In fact, MTs are plenty interesting, particularly to toxicologists concerned with evolutionary responses to chemical toxicity.

The MT protein is best understood in vertebrates, having been originally discovered in the 1950s as an odd cadmium-binding protein isolated from horse kidneys, although decades later it was identified in invertebrates collected from cadmium-contaminated sites, furthering its distinction as the "metal-binding" protein.[30] Setting MT apart from other proteins was a highly conserved functional region, rich with sulfur-containing and metal-binding cysteine amino acid residues. Yet, like shape-shifters with metallic hearts, MTs lack any structural allegiance, save their metal-binding regions. Further, MTs are able to associate with several metal ions at once—up to a dozen, depending on the species in which the protein is expressed, and the metal involved. These bound metal ions can account for up to 11% of the protein's weight.[31] It is as if the protein's main function is to provide life with sulfur-rich metal magnets (recall the sulfur-metal attraction), recreating an inner environment not unlike that existing before the GOE. This small protein (around seventy amino acids, in comparison to some CYPs that have upward of five hundred amino acids) displays affinity for both essential and nonessential metals, and, importantly, is inducible. Like catalase, possibly photolyase, and (as we will see) some CYPs, exposures to small amounts of metals, including cadmium, zinc, or copper, can increase the cell's production of MT.[32]

Based on the above, the role of MT seems straightforward. Metallothionein likely evolved to sequester and perhaps transport potentially toxic metals. Its function may have evolved roughly 2.4 billion years ago, when zinc, copper, and other metals were released by the GOE; or, if the zinc-world scenario holds, perhaps over a billion years earlier. But what of its affinity for cadmium? Our single-celled ancestors did not evolve on the edge of a battery factory or a foundry. What kind of selective pressure could a relatively rare metal like cadmium have exerted? And although their association with cadmium first revealed their existence, MTs also bind zinc, copper, and a slew of other metals, including lead, mercury, silver, gold, and bismuth.[33] Finally, MTs are documented to respond to an incredibly broad range of stimuli in addition to metals, including inflammatory agents, hormones, antibiotics, tumor promoters, and a range of proteins and chemicals produced during stress-inducing situations.[34] Could one of those responses be more closely related to its original function? Did it originate as a general stress-response protein with an odd affinity for metals? Or a metal-binding protein able to respond to other stress? When it comes to environmental stresses, particularly co-occurring stress, might trade-offs in terms of response time, energy required for protein production and control, and consequences of too narrow a response result in selection of a broadly responsive, generally nonspecific protein? Immune stress, for example, may alter the need for zinc *and* result in increased ROS generation at the same time. As such, a multifunctional protein could be beneficial.

Metallothionein's lack of a highly conserved primary structure makes tracing its evolutionary history difficult. In vertebrates alone, the numbers of genes coding for metallothioneins, and their specificity for certain metals, expression in tissues, regulatory control of inducibility, and amounts of protein expressed, vary across species.[35] All that remains for the molecular biologist bent on phylogenetic analysis, it seems, are the few retained characteristics, including its cysteine-rich regions, small size, and general (this may not always be the case) lack of histidine.[36] It's not much, but enough to indicate that MTs evolved from a single ancestral gene.[37] The wide array of subfamilies of this protein, even within a given species, indicates a long history of gene duplication events. Duplication, which is exactly what it sounds like, is an important mechanism of evolution (as discussed in chapter 2). When retained, the products of duplication events are termed "isoforms." While mice, men, and other mammals benefit from four isoforms in the metallothionein family (seventeen MT genes have been identified in the human genome, although not all have been characterized, and some are associated with "subtypes" of the major isoforms),[38] the most widely expressed associate with zinc and copper. Meanwhile other species may have fewer, and very different, isoforms.[39] One feature of duplicated genes is that either they may remain under the same regulatory control as their progenitor, or they may splice into a different chromosomal location, providing the means for differential control of the two isoforms.[40] It's easy to see how duplicate genes retaining full function would benefit a population challenged with increased exposure to metals. The process of gene duplication, as we will see repeatedly throughout this book, builds gene families. In the case of metallothionein, the result is a genealogical lineage that would be difficult to trace if it were not for retention of its most basic characteristics.[41]

Because they are copies, duplicate genes can more freely undergo change, providing an important route for protein evolution as well as creating a buffer from harmful mutations.[42] As with other instances of mutation, the impact can range from silencing the gene, to a compromised protein product, or, more rarely, production of a more effective protein.[43] In the case of metallothionein, changes in specific amino acids within and around the cysteine core appear to be the cause of varying susceptibility to metals in different species.[44] Fruit flies, for example, have four genes for metallothionein, encoding one protein with an affinity for Cu, another having Cd-binding properties, and two having unknown specificity.[45] Earthworms have three Cd-binding metallothioneins, one of which is inducible and another that is more prevalent in embryonic tissues versus adult tissues.[46] Curiously, despite these differences, *all* vertebrate and invertebrates, it seems, have least one Cd-binding, sometimes even Cd-specific, MT. Which brings us back to why Cd? That life has required and retained protection against oxygen and UV, as discussed in previous chapters, is to be expected. But cadmium? Given the rarity of cadmium in comparison to copper and zinc (at least post-oxygen), the ubiquity of a cadmium-binding protein is curious. Perhaps cadmium has some beneficial function that has yet to be discovered. Or perhaps metallothioneins have other functions not yet understood, and cadmium binding is incidental. Or perhaps the relatively low levels of naturally occurring cadmium routinely encountered by living things are sufficiently toxic to warrant dedicated protection. Maybe the role of metallothionein is and always has been to bind potentially toxic metals.

Recent studies of metallothioneins in ciliates, a ubiquitous group of single-celled eukaryotes, offer some clues. Like their multicellular eukaryotic cousins, ciliates also produce a Cd-binding metallothionein. But amino acid analysis suggests that the cadmium-binding isoform found in ciliates emerged *subsequent* to the ancestral form of metallothionein, which displays higher affinity for copper, zinc, or both.[47] As will be discussed in the following chapter, single-celled eukaryotes lived on the cusp of the GOE, and we know that with oxygen also came increases in zinc. Unlike catalase and photolyase, metallothionein has not *yet* been traced as far back as the last universal common ancestor, or LUCA (possibly leading us to question its role in the zinc world), and MTs have not yet been characterized in archaea, although some forms (quite different than those found in eukaryotes) have been identified in bacteria.[48] While LUCA appears to have encoded genes for metal-binding proteins (including those for iron, molybdenum, and magnesium), they are thought to have been a different class of metal-binding proteins, involved in metal transport and biosynthesis.[49] Pending further evidence, the original role of ancestral metallothionein seems to lie somewhere in the transition from anoxic to oxic conditions, during the transition from zinc scarcity to zinc abundance. In cyanobacteria (and a few other bacterial species), the *primary* function of the metallothionein may have been the maintenance of zinc concentrations, although they do also bind copper and cadmium.[50] Considering the natural history of both zinc and essential metals, this makes sense. In an environment where zinc, once rare, became increasingly available to cyanobacteria (ironically, a result of their talent for photosynthesis), preventing newly abundant zinc from replacing or otherwise interfering with an essential metal like iron might have been the difference between survival and extinction. Over time, zinc worked its way into living things. Proteins specializing in the binding of either zinc or cadmium, as their external availabilities changed, would provide an efficient route to metal homeostasis (in the case of zinc), and protection against toxicity (in the case of cadmium, replacing zinc is one of the primary mechanisms of its toxicity).

Even if life had originated in the zinc world, the environmental changes wrought by the GOE resulted in large-scale environmental changes, providing a strong argument for the selection of a simple sulfur-rich protein that controlled not only zinc, but eventually other metals as well. We might imagine that as life became more complex (and in some ways less flexible, as specific dependencies increased)— moving from sea to land and back, along with the dietary changes that go along with such movements—the ability to protect oneself from new metals, along with the ability to seek out and acquire others, might have become ever more important for survival. The duplication and selection of a simple metal-binding protein provided a starting point for the evolution of multiple forms of metal-binding capability, variously specializing in zinc, copper, and, not too much later, cadmium. In our own bodies, metallothioneins are most strongly expressed in tissues likely to encounter metal toxicants on their first pass through the body (e.g., the blood, liver, kidney, and intestines), and may play a role in diseases caused by metal imbalance.[51] Given their role in balancing and protecting against common and potentially toxic metals, we might think these proteins ought to be as essential as are some of the metals with which they interact. Curiously, mice lacking metallothioneins seem to suffer no ill, except for a much greater susceptibility to Cd-induced kidney toxicity,[52] suggesting a redundant mechanism for maintaining homeostasis of essential metals like Zn and Cu, but not for protection against toxics like cadmium. Redundancy for essential functions makes sense, and is part of many defensive systems, from UV to ROS to organic toxicants, as we will see later. Yet at some point, metallothionein appears to have evolved into an important part of *the* defense against Cd. We are exposed to cadmium primarily through our diet: grains, meats, and especially seafood all contain cadmium. More recently, humans and wildlife have become exposed to cadmium and other metals because of industrial activity,[53] and increased Cd exposure in human populations could be associated with an increasing prevalence of kidney damage.[54] It is not difficult to imagine how human metallothioneins' capacity to respond might become overwhelmed by our modern environment—which brings us to scientists with a penchant for worms.

Contemporary Evolution in a Metal World

Back in the mid-1980s, Jeffrey Levinton and his PhD student Paul Klerks, both at the State University of New York at Stony Brook, were two of those scientists who poked around highly contaminated sites seeking worms. In this case, it was Foundry Cove, an inlet along the Hudson River that served as the dumping ground for a string of battery companies, ending at the time with Marathon Battery Company. After more than twenty years of NiCd battery manufacture, the amount of cadmium in sediments of Foundry Cove was as high as 5% in some of the most contaminated regions. Yet despite such potentially toxic concentrations of cadmium and other waste metals, populations of freshwater worms (*Limnodrilus hoffmeisteri*) flourished, apparently unperturbed by cadmium. They appeared to be protected by a protein consistent with, but at the time not confirmed as, metallothionein.[55] That organisms could survive in such a contaminated environment was foreseeable. Other populations of invertebrates were known to hold their own in contaminated sites. But was this purely a case of individual resistance though induction of high levels of a metallothionein, or was it a heritable response, an evolutionary change, passed on from parent to offspring? And if heritable, how quickly could this happen?

At the time of their discovery, only one population of invertebrates, whose environment had been contaminated for centuries rather than decades, was confirmed to have undergone a genetic change in response to metals.[56] That a population of rapidly reproducing organisms, capable of producing large numbers of offspring, under a strong directional selection pressure such as cadmium, might evolve resistance was not particularly surprising (this will be discussed in detail in chapter 9). But how quickly could a species evolve? Addressing this question, Levinton and Klerks designed a series of studies involving worms from both the contaminated site and a reference site. They not only confirmed heritable resistance in *Limnodrilus*, but also discovered that it could happen in just 1–4 generations: an evolutionary blink of the eye. Theirs was one of the first studies demonstrating a genetic basis for rapid evolution of pollution resistance.

Levinton's focus on these populations did not stop there. He continued his studies over the next decade, as the site was cleaned up through the EPA's Superfund program. Surprisingly, as the site became clean the worms lost their resistance. Not only that, but again the change was rapid (over the nine-year cleanup, an estimated 9–18 generations of worms, at the very most). When resistance is energetically "expensive" and therefore presents an unfavorable trade-off, its loss over time might make some sense. A population might also "lose" a trait if there is a tendency for nonresistant populations to immigrate, providing an opportunity for gene flow and dilution. Based on earlier observations that worms from the contaminated site grew more slowly (although general fitness and the production of viable offspring did not seem affected), Levinton suggested that the loss was associated with energetic costs.[57] Testing this hypothesis required recreating a resistant population in the laboratory, and then exposing the worms to increasing amounts of cadmium. The artificially induced "natural selection" of worms from both Foundry Cove *and* its reference site yielded two resistant populations. It turned out these populations were no different from nonresistant populations in terms of fitness.[58] Loss of resistance through immigration, it appears, rather than energetic expense, caused the loss of resistance in Foundry Cove.[59] This was an important finding, since resistance *may* come with a cost in terms of fitness (discussed further in chapter 9), which in turn might influence how observations of resistance are interpreted.

Foundry Cove worms yielded another interesting observation, one likely relevant to the evolution of toxic responses in general. The population whose ancestors had once lived with cadmium developed resistance faster than the naive, reference population. This may have been a consequence of a key mechanism of rapid evolution: standing genetic variation. Populations adapt to environmental change in at least two ways, through new mutation or selection on preexisting mutations. Despite dilution through immigration, if Foundry Cove populations maintained alleles that afforded Cd resistance, they most likely would have been able to respond more rapidly with the return of Cd. Retention of once-useful gene alleles (variants) creates a pool of alleles likely to have been "tested" over time. If these alleles are selected and maintained at higher frequencies compared to new mutations, they may contribute to standing genetic variation and phenotypic plasticity.[60]

All living things require metals, and a few billion of years of exposure have provided us with the means to balance the need for these potentially toxic elements against metal overload. While the oxygenation of Earth's atmosphere dramatically altered the availability of metals, in the past few hundred years humans have initiated further changes. Metals, from mercury to lead to "rare earth" metals like yttrium and cerium, once locked up in earth's crust, are now more prevalent than ever. The effect of such rapid, increased bioavailability has been devastating (and obvious) in some cases: mercury poisoning in Minamata, Japan, or elevated lead levels caused by leaded gasoline, paint, and lead shot. Other effects have been subtler, including neurotoxicity caused by small concentrations of mercury acquired through the diet. Yet some populations, as discussed in this chapter, have evolved resistance to certain metals through the rapid evolution of protective proteins, the metallothioneins—proteins that, among other functions (many still unknown), bind metals and help maintain balance. As the trace-metal biochemist Peter Coyle and colleagues observed, "The extraordinary degree of conservation of the functional structure of MT across phyla suggests that most of its evolutionary shaping was complete hundreds of millions of years ago, with relatively minor further structural changes, fine-tuning the chemical adaptation to specific (external) environmental and metabolic (internal environment) requirements."[61] There is still much to be learned about MT: why some species develop metal resistance, while others remain exquisitely sensitive to metal toxicity, and the role that MTs play in the protection of different cells and tissues within and across various species. With additional research, we may move toward a more holistic understanding of the importance of metals for health and in causing toxicity.

PART 2

Plant and Animal

Then

Eukaryotes emerged with increasingly complex genomes, and acquired mitochondria, likely through association with prokaryotic life

Toxic Evolution

The eukaryotic genome was amenable to the formation of multicellular life, and to the formation of cancer. Though a later arrival, genes like p53 help suppress cancer; in almost all cases of human cancers, the p53 pathway is nonfunctional

Now

Cancer is as old as multicellular life, yet exposure to industrial contaminants may overwhelm life's capacity to suppress cancer

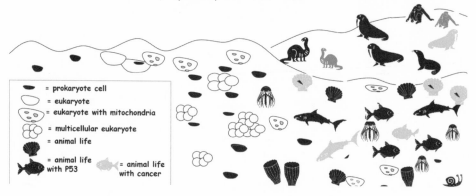

= prokaryote cell
= eukaryote
= eukaryote with mitochondria
= multicellular eukaryote
= animal life
= animal life with P53
= animal life with cancer

Evolution of cancer and P53 in metazoans.

Chapter 5

It Takes Two (or More) for the Cancer Tango

The cells acquire selective advantage because their accumulated set of mutations progressively unleash a pattern of inherent primal properties: persistent cloning . . . immortality . . . and a capacity for territorial expansion.

Mel Greaves

Only through integrated molecular, ecological and evolutionary analysis of cancer, at the somatic, population and macroevolutionary levels, will we come to understand and govern this unique disease.

Bernard Crespi and Kyle Summers

We have met the enemy and he is us.

Pogo *comic strip by Walt Kelly*

Approximately one and a half billion years ago, as individual cells began banding together, life diverged from its singular existence, laying the groundwork for the incredible animal life we behold every day. A spider tends to her web. A bird calls from the tree. My fingers tap a thought on the keyboard. But as life gained the capacity to silence one gene or turn on another, thereby differentiating into specialized cells and eventually into basic body parts, it also gained its own worst enemy—itself. Cancer is the renegade cell released from life's rules of order. It is rapid evolution at its best and worst. Its origins are as old, and perhaps older, than the first "true animals," and it is, by most accounts, a greater threat to the human population today than at any other point in our modern history. It is also a disease of multicellular life. As with any social organization (and multicellular bodies bear some semblance of organization), there are collaborators, independents, and renegades. So it should be no surprise that once organized, life also became host to the independents and renegades—and the battle between self and altered self, or cancer, has raged ever since.

Prior to this chapter, we focused on life's interaction with a set of physical and chemical toxicants—UVB, oxygen, and metals—and the history of the protective genes and enzymes that subsequently evolved. We now consider the evolution of a defensive system that protects against the consequences of life's *own* processes gone wrong, whether initiated because of spontaneous mutation, or helped along by physical or chemical mutagens. Of the many different responses to cancerous cells, one defensive mechanism in particular stands out, the p53 tumor suppressor protein. While the preceding chapters emphasized toxic defense mechanisms originating in our unicellular prokaryotic ancestors, here, and for the remaining chapters of this book, we enter the realm of eukaryotes, where another node in the defense network, the p53 gene family, emerges. Beginning with the eukaryotic cell, we trace the origins of the p53 gene family from its earliest appearance to one of its present roles—cancer suppression.

A Complex Cell

Because they are the cells on which complex life is built, there is a great deal of interest in the eukaryotic family tree. It is an evolutionary history stretching back at least 1.5 billion years and very likely earlier. We've all learned in basic biology that eukaryotes are fundamentally different from prokaryotes in several ways, including their membrane-bound nucleus, compartmentalized internal cellular environment, and cytoskeleton. Yet how and when these relatively advanced cells split from their prokaryotic ancestors remains a mystery. Advances in gene sequencing and data analysis have only served to raise more questions about their evolutionary relationship—particularly when it comes to the mitochondria, the so-called powerhouses of the cell. Those membrane-bound organelles are defined by their own genetic material (referred to as MtDNA) and are a characteristic feature of most eukaryotes.

Likely because of an encounter between two ancient cells, mitochondria eventually became permanent and essential to eukaryotes.[1] Many have pondered the initial nature of this relationship (predator and prey, or host and parasite?); the nature of the two cell types (two prokaryotes, or prokaryote and eukaryote?); and any potential costs and benefits of the relationship to the two participating cells (did the host cell gain a source of energy, while the "invading" cell was well fed and protected?).[2] These questions concerning eukaryotic evolutionary history have yet to be answered and may seem to divert us from our focus on toxic defense—unless one early benefit of mitochondria *was* detoxification. In his book *Oxygen*, Nick Lane considers the oxygen defense alternative for mitochondria. Based on observations of mixed communities of aerobes and anaerobes, Lane asks what if instead of immediately benefitting from the extra energy produced by invading cells, the internal aerobic cell instead provided an advantage to its host by detoxifying oxygen? By converting oxygen to water, as would happen during aerobic respiration, these organisms may have provided their hosts with benefits, while the host kept these living oxygen filters fed and sheltered. Additional benefits—particularly the export of the high-energy ATP, which required transport of the bulky and relatively unstable molecule across the mitochondrial membrane—may have evolved later.[3] As the earth's oxygen concentrations rose, this newfound tolerance may have allowed for the expansion of eukaryotes into new territories, while the additional energy may have encouraged the expansion of their genetic repertoire.

No matter the original relationship, mitochondria and a dependency on oxygen are now firmly entrenched characteristics of the great majority of eukaryotes, including the trillions of eukaryotic cells on which our lives depend.[4] And the incorporation of mitochondria is now recognized as one of life's major evolutionary milestones, along with the development of a multicellular lifestyle.[5]

Life beyond the Singles Bar

How and when eukaryotes first became organized into multicellular life is yet another mystery. Did toxic oxygen play a role, perhaps by causing cells to clump together for protection, with outer layers protecting the inner layers? Or did cells, in an attempt to detoxify oxygen, sequester it (à la metallothionein) into biomolecules like cholesterol and other oxygen-rich molecules, now a distinctive feature of most eukaryotic cell membranes?[6] Maybe, after hundreds of millions of years of adaptive selection and evolution, an increasingly diverse genome provided the basic building materials for complex life? Perhaps. But the evolutionary geneticist Michael Lynch suggests that rather than relying on adaptive selection, the complex gene structure of eukaryotes might instead have evolved as a result of nonadaptive processes, a consequence of population size and the machinations of population genetics.[7] One characteristic of eukaryotes is their larger size in comparison to prokaryotes. And larger organisms tend toward smaller populations (the smaller the population size, the greater ability to propagate a mutation). Given this scenario, reasons Lynch, population genetics, including genetic drift and random neutral mutation, might have driven the basic changes in the eukaryotic genome, which in turn provided the components on which natural selection could then build. In his 2006 publication on *The Origins of Eukaryotic Gene Structure*, Lynch does not hold back on the omission of population genetics when reconstructing evolutionary histories, writing, "Because natural selection is just one of several forces contributing to the evolutionary process, an uncritical reliance on adaptive Darwinian mechanisms to explain all aspects of organismal diversity is not greatly different than invoking an intelligent designer."[8] The role of population genetics is often neglected, yet as Lynch points out, it likely influences the evolution of many biological systems. Although the topic cannot be done justice here, those interested might begin with Lynch's book *The Origins of Genome Architecture*.

That said, not all would agree that life simply drifted into complexity. Controversy abounds when reconstructing evolutionary histories. Pointing to the likely abundance of small prokaryotic populations over the eons and the rare emergence of eukaryotes (as far as anyone knows, eukaryotes evolved from prokaryotes only once in four billion years), Nick Lane and William Martin offer another route to genome complexity. Based on their analysis of the energetics of increasingly complex genomes, they suggest that while the "prokaryotic genome size is constrained by bioenergetics," the acquisition of energy-producing mitochondria (once ATP export evolved in earnest) provided cells with a six-figure energy surplus.[9] This energy, suggests Lane and Martin, could then be allocated to the expensive business of diversifying a cell's genetic complement and protein production—thereby increasing genetic complexity.

Whatever the driving force, whether population genetics, energetics, oxygen toxicity, or some combination of all the above, the increased complexity of both cellular architecture and the gene allowed for new features now characteristic of eukaryotes. These included improved facility for cell-to-cell communication and coordination, adherence between cells, and, counterintuitively, the capacity for a cell to die on "command." Programmed cell death, or apoptosis, is one of the essential processes that shape multicellular life. Consider the transition from a fetus's webbed hands to the delicate fingers of a newborn. That the hands of a newborn do not resemble the wings of a duck or the fins of a fish can be attributed, in part, to apoptosis. Apoptosis also helps keep cells with irreparable DNA damage in check. Like a spy with a suicide capsule tucked away in her tooth, cells that may be dangerous to the whole (because they are compromised beyond repair) are imbued with the capacity to commit suicide. Apoptosis is so important to life that many different genes are capable of initiating the process—including the tumor suppressor p53. Directing assisted cell suicide in DNA damaged cells, it seems, is one of p53's premier roles, as we will see.

Apoptosis, along with communication, adherence, and coordination, are all important components of the multicellular tool kit, tools that may have allowed single eukaryotes to build multicellular homes for themselves.[10] As multicellular eukaryotes or metazoans, we share a large portion of our eukaryotic genome with all animal life, from sponges to fruit flies to dogs.[11] While many eukaryotes diverged, undergoing a great deal of evolutionary change over the years, others seem to have changed little. Those of us who reach for the "all-natural" sea sponge when we shower are reaching back at least six hundred million years to one of the few tangible reminders of our predecessors. With about a dozen somatic or body cell types, in contrast to upward of two hundred or so cell types that make up our own bodies, sponges are among the oldest living representatives of multicellular life.[12] While sponges may be relatively simple, their genome includes the most basic components of the multicellular tool kit, those hallmarks of metazoan life that differentiate the inner workings of multicellular life from our single-celled ancestors.[13] Which brings us back to cancer and p53, for, ironically, this tool kit is partially responsible for the evolution of cancerous cells. The first traces of cancer genes and of the p53 gene family are found within the sponge genome, and those of other early invertebrates like the sea anemone, dating the roots of the disease (and perhaps its defense) back to well over half a billion years. Yet, for reasons that are not well understood, cancer, if defined as an invasive growth of mutated or transformed cells, is rare in many of our invertebrate relatives.

It is not difficult to imagine that with the greater number and diversity of cells comes increased potential for the emergence of renegades. Life has retained at least two sets of genes that, at the very least, help maintain crowd control. Although prokaryotes need only maintain the genomic integrity of their singular selves, and do so by relying on ancestral genes involved in processes like DNA repair and maintenance (the so-called caretaker genes), multicellular eukaryotes must maintain genomic integrity *and* ensure that any cells managing to outwit the caretakers are also kept in check. Genes involved in that process are referred to as "gatekeepers" and are dedicated to controlling cellular functions characteristic of multicellular life, including cell-to-cell communication, growth regulation, and death. Any time a cell breaks free from these controls, it has the potential to become a renegade. In a sense, cancer may be defined as the dysfunction of a sufficient number of both caretaker and gatekeeper genes, leading to uncontrolled growth. Although it is inevitable that dysfunctional caretaker genes will arise in prokaryotes, the association of gatekeepers primarily (although not exclusively)[14] with multicellular life, and *their* dysfunction with cancer, perhaps places cancer's roots in the realm of multicellular organisms.[15]

Indivisible, with Cancer for All

It was 1982 and I was in need of a senior college thesis. Curious about chemical contaminants, I devised a simple study to evaluate the impacts of pollution on blue mussels in Hull Bay, a small inlet not far from one of the most contaminated waterways in the nation, Boston Harbor. Hypothesizing that mussels from the bay, weakened by industrial and domestic pollution, would be more susceptible to parasites than would mussels from the reference site, I enlisted a local marine pathologist, Bob Hillman, for advice. After weeks of examining the pink- and blue-stained sections, identifying the telltale swirl of a parasite from normal tissue was easy. Yet one sample was different: the cells were disorganized and stained more intensely. A cursory pass under Bob's scope prompted him to suggest a tumor. Ignorant of the implications, and eager to finish, the thesis was completed with merely a mention of a possible, but unconfirmed, tumor. Several years later Bob inquired about the slides. Those disorganized cells, he'd suggested, might in fact have been one of the first findings of a cancerous tumor in Boston Harbor mussels (unfortunately, the tissue samples were long gone by then). A short time later, two different studies confirmed the formation of cancerous tumors in bivalves from Boston Harbor and other locations, while laboratory studies linked sediments laden with polyaromatic hydrocarbons (PAH) with tumors in bivalves. This was a new finding for toxicology, but nothing new to life.

Afflicting one in three people over a lifetime, cancer is also a disease of clams, fish, and frogs—even fossilized dinosaurs show indications of cancer.[16] Documented in humans thousands of years ago and named by the Greeks, cancer is a collection of diseases with a long recorded history. Yet it remains poorly understood—perhaps until now.[17] Cancer is both a disease of renegade cells, governed by their own damaged genome, and a disease of the body's *social organization*.[18] It is also the price multicellular life pays for the ability to function and evolve.[19]

As a disease requiring genetic mutation—as suggested by the somatic mutation theory of carcinogenesis—if all mutations or mistakes in DNA replication were correctly repaired, and *if* cancer truly requires a somatic mutation, cancer would not exist. Nor would we have evolved to ponder cancer, toxicology, or damage control. (Of course, there are alternative theories for the origin of cancer, including one focused on tissue organization rather than individual cells.)[20] Evolution begins with mutation, yet some measure of DNA repair is required to provide for genomic stability, whether for a single-celled species or for complex life. The example of DNA photolyase, the enzyme responsible for repairing DNA damage caused by UVB (highlighted in chapter 2), reveals another facet of the inevitability of cancer. In addition to errors inherent to DNA replication, chromosomal material is vulnerable to a broad range of environmental influences, both internally and externally. Besides UVB, DNA integrity is at the mercy of reactive oxygen species, ionizing radiation, viruses, and a whole host of naturally occurring mutagens (PAHs and mold-produced aflatoxin, for example) and industrially released mutagens (including PAHs from fossil fuel combustion, PCBs, and benzene). Any physical and chemical toxicants, individually or in combination, can start a cell on its way down the path to cancer—and all too often does.

Genetic mutation, however, is only a first step in the process. Cancer is also a disease of cell regulation. In their highly influential paper "The Hallmarks of Cancer," the molecular oncologists Douglas Hanahan and Robert Weinberg identify six alterations essential for the evolution of cancer in humans: self-sufficiency in growth signals, insensitivity to growth-inhibitory (antigrowth) signals, evasion of programmed cell death (apoptosis), limitless replicative potential, sustained angiogenesis, and tissue invasion and metastasis. Together, write the pair, these alterations "represent the successful breaching of an anticancer mechanism hardwired into cells and tissues."[21] Cancer makes short work of the multicellular tool kit, as the confluence of cell reproduction, growth, and control enables both multicellular life and cancer.

As I ponder the tens of trillions of somatic cells in my body, the tens of thousands of genes per cell, and the numerous opportunities for mutation, I wonder if and when cancer might strike. With good fortune and good genes, maybe it never will. Or perhaps it already has, and I am unaware. Given the opportunity, it is a wonder that cancerous growths are not more pervasive. How complex life avoided this fate may reside in the mechanics of cancer, the body's defenses, and the fact that all genes do not contribute equally to the evolution of cancer. Recent efforts to characterize the genomes of cancer cells are providing insight into the specific genes that, when mutated, are most likely to contribute to cancer in humans. Currently the proportion of genes associated with somatic cancer (and the great majority of cancer occurs in somatic, as opposed to germ, cells) represent just 1.6% of the human genome—a small proportion, although a figure that is likely to increase with increased sequencing efforts.[22] When mutation hits the DNA repair genes, a door may more easily be opened for the renegades. For example, although we lack the DNA photolyase repair gene, we are somewhat protected from UVB damage by the far more complex nucleotide excision repair (NER) system (discussed in chapter 2). Yet defects in any one of a number of different genes involved in the NER pathway underlie a condition known as xeroderma pigmentosum—the inability to repair UVB-induced DNA damage. It is also associated with a thousandfold increase in the risk of UVB-induced skin cancer in humans, a harsh reminder of the consequences of inadequate DNA repair.

Besides being relegated to a relatively small set of genes, cancer often requires multiple mutations involving different categories of genes, collectively referred to as cancer-susceptibility genes.[23] Mutations in the caretakers and the gatekeepers are just bricks along the road. Yet the road to cancer is not a straightaway. Rather, it is more like one that might appear through Lewis Carroll's looking glass. It is a road that twists and turns, with roadblocks and loopholes. It is a road where one turn influences another, and any action can set off cascades of unintended consequences. While inactivation of a single caretaker gene may be insufficient to cause cancer, the genetic instability that follows opens the door for mutations in other genes, including the gatekeepers, potentially leading to uncontrolled cell growth. Like a ball released at the top of a hill, once cancer is on its way, chances are it will continue to roll—although the speed at which it rolls and the route it takes depend on everything from age, to nutrients, to environmental stressors like too much ozone, changes in oxygen levels, and exposure to ubiquitous combustion products like PAHs.

However (and perhaps fortunately), even for cancerous cells, too much of a good thing can be disastrous—and increased mutation rates as a result of defective caretakers may not always benefit transformed cells. The evolutionary biologists Barnard Crespi and Kyle Summers explore this conundrum in their 2005 review, "Evolutionary Biology of Cancer." While a mutagenic environment may be favorable for cancer initiation, it is also a harsh environment for survival. Crespi and Summers suggest that a cell must balance the need for repair with the cost of repair (a mutation rate that spirals upward will require more repair, even in a cancer cell). If too expensive, then some cells (or cancers) may optimize and limit mutation rates, and this balancing act, write Crespi and Summers, "has crucial implications for cancer therapy, because many chemotherapeutic agents are themselves selective mutagens that might promote the instability that ultimately renders them ineffective."[24] Additionally, as a potentially deadly example of rapid evolution, some cancer cells evolve resistance to one cancer drug after another. Genomics, combined with an improved understanding of evolutionary processes, is just beginning to provide the insight needed to outwit those rogue cells within us.

While research into the cancer genome, and the identification of caretaker, gatekeeper, and landscaper genes (which provide support for growing tissues, like those governing blood vessel growth), continues at breakneck pace, one gene in particular has stolen the show, at least for now.[25] Deriving its name from the molecular weight of the protein for which it codes, and discovered roughly thirty years ago, p53 is referenced in more than seventy thousand published articles. With an altered form of this gene identified in at least half of all human somatic cell tumors sampled from cancer patients, and alterations in the p53 pathway in almost all cancerous tissues, it has certainly earned its celebrity.[26] As we will see below, while p53's role in life is not always clear, there is good reason it is considered one of life's most important cancer-defense genes.

p53: A Family History of Death and Destruction

With multicellular life came lifestyle changes, including longer lives, larger bodies (in general), and increased opportunity for DNA damage from exposure to both internally produced toxic by-products like reactive oxygen species, and external mutagens like certain metals, PAHs, and UV light. Yet as life's internal environment became more favorable toward the evolution of cancerous cells, it also promoted the evolution of cancer-prevention tactics. One outcome may have been the emergence of p53—one of a trio of genes including p63 and p73—as a key gene in cancer control.[27] These are all multipurpose genes, and while the roles of their gene products are not fully understood, they do have one thing in common—they induce apoptosis in cells with DNA damage. Although the gene family's history reaches back to unicellular eukaryotes, the emergence of a functional gene for p53 (as of yet missing in early metazoans and invertebrates) is a relatively recent event.

The first inklings of the p53-gene superfamily appear in choanozoans, the single-celled eukaryotic predecessors of multicellular life.[28] These genes date back at least six hundred million years to a time when metazoans and choanozoans began to diverge. The representation of the p53 superfamily in all metazoans, from invertebrates to vertebrates, suggests that there was likely some benefit to retaining these genes.[29] Given their ability to induce suicide in damaged cells, their utility is not difficult to imagine—unless this took place in single-celled life. Although retention of a suicide gene in single-celled organisms might spawn some interesting philosophical discussions, there are other roles this gene may have played, including delaying cell reproduction in damaged cells, providing the cell with an opportunity for repair—a function of p53, and possibly its genetic ancestors as well. p53 itself is the youngest known member of the gene superfamily. Its first ancestor appears to have been a gene resembling a hybrid version of p53's elder siblings, p63 and p73 (referred to here as the hybrid). This hybrid may eventually have provided the building blocks for p53. So far, the hybrid's function in living representatives of early metazoans (like sea anemones) seems not too far afield from that of modern members of the p53 gene family, particularly the older members.

The starlet sea anemone, despite its simple nature and few cell types, reproduces both sexually and by fission—pinching off a fragment, which then regenerates. It is among the oldest members of the animal line, and the most ancient metazoan in which the hybrid gene has not only been identified, but also is functionally characterized. When exposed in the laboratory to UV light, the anemone hybrid induces apoptosis in damaged gametes.[30] In other words, the hybrid appears to protect the germ-line integrity of the species. The starlet lives in brackish waters at or above the high tide mark, unprotected from the sun's glare, creating significant challenges to maintaining germ-line fidelity. In addition to spontaneous mutations, it must contend with plenty of UV light. Given the sea sponges' habitat and multicellular body, it is not surprising that recent genomic analysis also suggests the presence of p53 gene family components in the sponge—perhaps protecting the germ-line integrity of this early animal as well.[31]

Yet p53 as we know it today, rather than protecting the germ line, is a tumor suppressor in *somatic* cells. How and when did this happen? We can imagine that as life continued to diversify, bodies (or soma) of some species became more prominent features, in contrast to the germ cells. As life span increased, animal bodies lasted longer—as did the opportunity for spontaneous mutation and exposure to physical, chemical, and even biological mutagens. Are all species equally susceptible to cancer? And can we learn something about cancer and p53 from its distribution across species? Among the many differences across species is the trajectory from embryo to adult. While some, like humans, continue to grow and regenerate throughout a lifetime (at least some of our tissues), other species do not. Take fruit flies and nematodes. The adult forms of these species are referred to as postmitotic. That is, their adult somatic cells no longer divide. This means that although they may accumulate DNA damage, they are essentially a genetic dead end—adult somatic mutations are not perpetuated, placing these species, perhaps, in a lower risk category for the emergence of somatic cell cancers.[32] While these species do express a p53-*like* gene, it tends to be more like the older p63 and p73 rather than p53 (and curiously, at least in fruit flies, reducing the activity of this gene in adult brain cells increases longevity).[33] But what of other species whose adult cells do continue to grow, adding and replacing tissues throughout adulthood? Consider regeneration in the starfish— the bane of fishermen, who tore the voracious shellfish predators apart, only to find they'd regenerated their missing bits and multiplied. Are they susceptible to somatic cell cancers, and does some member of the p53 gene family protect these species? Or what about bivalves, some of which can grow to the size of a large child and live for hundreds of years? Here at least are the first indications of a functional p53 protein, and perhaps an intriguing example of life beyond the genome. Although apparently lacking a dedicated p53 gene, some bivalves may receive somatic protection from a p53-like protein. Shellfish, including the soft-shell clam (*Mya arenaria*) and the blue mussel, may produce this protein by patching together pieces of the p63/p73 hybrid. The resulting protein even seems to behave more like p53 than its hybrid ancestor, making an appearance in cancerous somatic cells, which suggests some role in cancer protection.[34]

The first known appearance of a gene dedicated to the production of the p53 protein comes with the vertebrates. Perhaps their increasingly complex body structures, longer life spans, and delayed maturation and reproductive cycles provided favorable conditions not only for cancer but also for selection of p53. The earliest known vertebrates to host a gene for p53 are the cartilaginous fish, including rays and sharks (which are, contrary to previous beliefs, susceptible to cancer). These species also retain a p63/p73 hybrid, with the divergence into p63, p73, and p53 first appearing in bony fish.[35] The emergence of these genes from a single ancestral hybrid, like the metallothioneins and other protein families discussed in this book, provides yet another example of gene duplication and subsequent modification. These three genes have not only been conserved throughout the vertebrate lineage, but they have also retained their capacity for initiating apoptosis. Perhaps overshadowed (at least for now) by their younger sibling's role in somatic cell cancers, p63 or p73 remain largely mysterious. Yet there is growing evidence that their gene products protect the germ line, in addition to their other roles related to development. The p63 gene, for example, is required for skin and limb growth (mutations in human p63 result in cleft palates and other epithelial-related abnormalities), while p73 seems to be involved in neuron and immune cell development in addition to germ-line protection. And recent studies also suggest a role for both p63 and p73 in reining in cancerous cells.[36] But, for now, p53 is still the front-runner for cancer research.

Apoptosis, as directed through genes like the p53 family (there are also many other genes capable of invoking apoptosis), is like a constant gardener continually pruning back cells, as body parts are shaped in the developing fetus, or as adolescent brains mature, or as uterine linings are shed with each passing menstrual cycle. Apoptosis is also critical for proper immune function, from preventing autoimmune responses to inducing death in virally infected cells. Perhaps as vertebrate bodies grew in complexity and size, as their germs cells became increasingly sequestered from the outside environment, and as long-lived somatic bodies bore the brunt of environmental mutagens like UVB, oxygen, and metals, life became more skilled at combating its own worst enemy: itself.

Avoiding a Death Sentence: When Cancer Strikes

To think that any one of our cells, with the flip of a few genetic switches, could suddenly take its own life is an odd thought. But cells in our bodies are constantly committing suicide by the billions. That we don't wither and die the minute we are born (well, we do—it's called aging, another outcome of apoptosis) is the result of a complex network of genes and metabolic processes that control genes capable of inducing apoptosis. These networks respond to both internal and external signals, ranging from oxygen concentrations to metabolic stress, immune responses, and the everyday workings of our cells. Under normal conditions, signals in this network allow just enough p53 for "housekeeping," including maintenance of stem cells, temporary arrest of cell growth that allows time for repair, and, in aged or irreparably damaged cells, apoptosis. Considering p53's role in the execution of damaged cells en route to uncontrolled growth and cancer, it makes sense that cancer cells might skirt surveillance and destruction by selecting against functional p53 genes.[37] Increasingly, mutations to both caretaker and gatekeeper genes are being identified in cancerous tissues. In half of cancer cases, there is a mutation in the p53 gene, the causes and consequences of which are of utmost interest to cancer researchers.

So here we have a single gene producing a protein that is involved in some way in cancer suppression. Now is a good time to broaden our horizon and consider some internal and external factors that might influence the activity of its gene product. Recall from chapter 2 that the activity of a protein depends on folding and maintaining its shape. Additionally, one-third of our proteins are metalloproteins, requiring the presence of specific metals, like zinc, for proper functioning—and this includes p53. The single zinc ion located near p53's highly conserved DNA binding region is credited with maintaining the protein's functional shape. Without zinc, p53 fails to function. This requirement might explain at least two different routes to p53 dysfunction, both of which interest cancer biologists and chemotherapy developers. A mutation that reduces the binding capacity for zinc will affect p53's capacity for inducing apoptosis (this has been observed), and changes in zinc concentrations may also influence p53's ability to function (also observed). How does this play out in real life? Mutations in the zinc-binding regions of some cancer patients have been associated with poorer prognosis, although associations between p53, cancer, and zinc concentrations have yet to be observed.[38]

We know that metals are not only essential but also potentially toxic, and therefore must be tightly controlled. This brings us to metallothioneins (MT) and the possibility of a role for these metal-binding proteins in mediating some part of the cellular unrest that is cancer. As discussed in the previous chapter, MTs control metal concentrations in cells, including zinc. So what of the relationship between p53, zinc, MT, and cancer—or other metals for that matter? Although several studies measuring MT concentrations in cancer cells have associated increased concentrations of MT with a prognosis for poorer outcomes for a range of cancer types, the causal relationship between cancer and MT has yet to be determined.[39] But these findings have implications for chemotherapy (which often utilizes metals) and the evolution of drug-resistant cancer cells, in addition to providing some fodder for speculation about the interaction between exposures to metals and cancer. (Could metals contribute to the induction of cancer by altering the equilibrium of essential metals—particularly those involved in keeping cancer in check?) Whatever the relationship, the interaction between zinc, p53, and MT provides us with just a glimpse into both the complex network required to maintain balance of a single component in any given cell, and the consequences of disturbing this balance.

One additional observation of p53's role in life bears some consideration. No doubt a key factor in safeguarding vertebrate and possibly some invertebrate animals from cancer, p53 cuts both ways. Its destructive actions may be indiscriminate. That is, p53's activity is imperfect and sometimes makes no distinction between those cells with cancerous potential, and those which, given time and resources, could possibly have been repaired and returned to service. Considering this duality and its evolutionary history, the cancer biologists Melissa Junttila and Gerard Evan write, "p53-mediated DNA damage seems a blunt and inaccurate tool for suppressing tumors and, at worst, a dispensable relic of the checkered evolutionary legacy of p53."[40]

This trade-off between multifunctionality with problematic side effects, and specificity with limited utility is a common theme throughout this book. Metallothioneins can sop up harmful metals and inadvertently sequester essential metals. The CYP enzymatic pathway discussed in the next chapter, one of the premier detoxification systems protecting us from myriad environmental toxicants, defuses and activates toxic chemicals. The healing properties of the inflammatory response often come after damage caused by the release of toxic chemicals, including hydrogen peroxide. Perhaps these trade-offs simply reflect the outcome of a long, blind, evolutionary process.

As Junttila and Evan so eloquently point out, "Like some pervasive computer operating systems, p53 is an archetypical example of the unintelligent design and compromise that is inherent in evolution—a multifunctional, multipurpose transcriptional coordinator that has only lately been retasked to the job of tumor suppression in large, long-lived organisms. . . . At the end of the day p53, together with all our other suppressor mechanisms, fails half of humanity."[41] Today, more than in the past, this failure may not purely be the result of a gene system that fails under "natural" conditions, but rather under "unnatural conditions." Between dietary changes and the myriad industrial chemicals released into the earth's environments, we have imposed new external and internal environmental conditions on networks tuned to a different set of conditions. It's another example of an evolutionary mismatch—in this case, one that may benefit those renegades among us.

Then
A single-celled algae colonizes dry land, paving the way for subsequent evolution of land plants

Toxic Evolution
These immobile plants become targets for predators. In their defense they produce secondary metabolites; in response, animals evolve a diversity of detoxification enzymes and the plant-animal warfare begins

Now
Detoxification enzymes metabolize toxic chemicals and drugs, and are in part responsible for interspecies and interindividual differences in chemical susceptibility

= plants with CYP7s and other CYPs

= insects with CYP6 and other CYPs

= animals with CYP1, CYP3, CYP2 and other CYPs

Plants and the evolution of CYP detoxification enzymes.

Chapter 6

Chemical Warfare

Life could get along without animals and without fungi. But abolish the plants, and life would rapidly cease.

Richard Dawkins

P450 substrates in the past 1,200 million years then included sterols . . . endogenous metabolites, environmental chemicals, and plant metabolites. Since drugs are usually plant metabolites or derived from plant metabolites, the evolution of different P450 enzymes becomes central to the field of pharmacogenetics.

Daniel Nebert and Mathew Dieter

But the most famous plant antidote is that of Mithridates, which that king is said to have taken daily and by it to have rendered his body safe against danger from poison.

Aulus Cornelius Celsus

In comparison to invertebrates, microbes, and, most important, plants, humans are amateurs when it comes to chemical production and chemical warfare. While we refer to the first chemical revolution as a period beginning in the eighteenth century, a time when chemistry was demystified and humans began producing and creating synthetic and organic chemicals, we are hundreds of millions of years behind nature's chemical revolution. This revolution was driven in large part by plants.

As life emerged from its ancestral waters and took to the land, the animals roamed, but plants became anchored. So, unlike their highly mobile predators, plants needed protection.[1] Evolution provided a solution through the innovation of defensive, toxic chemicals. Some plant chemicals—like ricin in castor bean seed coatings, or atropine in the berries of deadly nightshade—are acutely toxic. Others, like the clover estrogens responsible for causing stillbirths and sterilization in Australian sheep, work in more subtle ways. Then there are those that fill our medicine cabinets and relieve our pain, stave off our cancers, and control our moods. We have not only evolved ways to defuse plant toxins, but have also found ways to benefit from nature's chemistry.

While the simplest defense we have against plant toxins is avoidance, plants form the base of almost all of the earth's food webs. Omitting plants from our diets just isn't an option. And so, just as nations warring over scarce resources modify their tactical response to new defenses, so too does life—albeit through the process of evolution. Animals evolved methods to avoid, excrete, and detoxify plants' poisons. While we may be relative neophytes when it comes to chemical *production* (although advances in nanotechnology may change that), when it comes to defense, humans and other complex animals possess a highly evolved, networked system of enzymes that detoxify. Of these, the cytochrome P450 (CYP) enzyme superfamily stands out as one of the most comprehensive chemical defensive systems known in animals—and plants provided much of the environmental selection pressure responsible for the evolution of large branches of the CYP family tree. In other words, we have plants to thank for our ability to readily metabolize and detoxify many dozens, and perhaps hundreds, of plant chemicals, pharmaceuticals, and even synthetic chemicals.

The number and diversity of CYPs is so sprawling that some have suggested that it be thought of in terms of its own genome, the CYPome.[2] As heirs to this system, humans metabolize and detoxify a dizzying array of chemicals, particularly those plant chemicals that have been part of our diet for millennia. On the flip side, for thousands of years, humans have also relied on plant chemicals for herbal remedies and medicines, and there is evidence that we may not be the only animal species to do so.[3] By some estimates, at least 25% of all modern pharmaceuticals are, or were at some point, derived from plants, and many of these are metabolized by CYPs. Because of its role in detoxification and metabolism, familiarity with the evolutionary history of the CYP system can go a long way toward understanding and predicting how humans and other species will respond to environmental chemicals. The comingling of land plants and animals appears to be the root cause of one of the planet's first chemical wars: a conflict originating 300–500 million years ago, and which continues to this day. So, before focusing on the CYPs, some consideration of the history of terrestrial plants and land animals is in order.

The Combatants: Plants and Animals

Just as we share a single-celled eukaryotic ancestor with sponges, starfish, and marmosets, land plants also share a single-celled ancestor. Most likely this was a green algae that 470–500 million years ago managed to survive and reproduce on land.[4] The descendants of this species converted the earth from a landscape dominated by minerals to one sheathed in green—one rich in organic, carbon-containing molecules—and transformed its atmosphere and the cycling of carbon. Green plants inspired carbon dioxide and exhaled oxygen while transforming the sun's energy into the sugars and starches coveted by animal life. From algae came mosses, ferns, conifers, and flowering plants—verdant life covering the earth and providing a vegetarian smorgasbord for animals to feast on.

Following the evolution of photosynthesis in bacteria, plants and algae have served as the foundation for almost all food webs on the planet. Yet for reasons not fully understood roughly 475 million years ago, plants expanded beyond their aquatic, nomadic lifestyle, settling along shorelines and eventually anchoring themselves in place, setting roots into the earth.[5] This change likely provided greater access to certain minerals and atmospheric carbon, greater territorial coverage as they moved farther from water sources, and an increased opportunity to soak up the sun's rays as they spread their branches and leaves far and wide.

But being grounded has its pitfalls. Although all plants (whether terrestrial or marine) are subject to predation, rooted land plants became not only easy targets but also locally dependable food sources. A perfect setup, it would seem, for rapid extinction, if it were not for two things: (1) their capacity for defense; and (2) their ability to benefit from predation.[6] While the beneficial relationship (including pollination and spreading of seeds) between plants and animals is interesting, and not altogether irrelevant, it is well beyond the scope of this book. The *adversarial* relationship with predators, however, fits right in.

Just as plants began colonizing solid ground, so too did animals. There is some agreement that the earliest animals to make landfall were likely arthropods, a category of invertebrates including insects, crustaceans, and arachnids. Some of these creatures left fossilized tracks estimated to be at least 480 million years old.[7] The two main groups of animals, protostomes (which includes sponges, mollusks, and arthropods) and deuterostomes (which includes chordates, vertebrates, and some closely related invertebrates) diverged about a billion years ago. Arthropods most likely encountered the vengeance of terrestrial plants well before vertebrates did. When tracing the evolution of CYPs, this is a particularly significant split.

With the earliest indications of true land plants currently estimated to be almost concurrent with the arrival of arthropods (give or take a few million years), the race for landfall seems to have been close, in contrast to old notions of animal life slogging its way onto a lush planet, in search of food. Additionally, rather than a race to the salad bar, fossil evidence suggests that the first terrestrial arthropods (and vertebrates) more likely fed on each other, detritus, and perhaps fungi, rather than on living plants.[8] This means that for one hundred million years or so, terrestrial Earth belonged to the plants, arthropods, and any single-celled creatures that colonized the relatively dry ecosystem.[9] All the while, aquatic vertebrates continued diversifying from their early chordate relatives (exemplified today by the sea squirt) to some version of a bony and eventually lobed-finned fish.[10] It was not until late in the Paleozoic, sometime between three hundred and four hundred million years ago, that our lobe-finned ancestors first overcame the problems of gravity, desiccation, and extraction of oxygen from air rather than water, and made a life on land.

What first drew vertebrates to land is unclear. Perhaps it was better hunting, as early tetrapods were carnivorous.[11] Or perhaps land provided a respite from other aquatic predators or the low-oxygen conditions in their aquatic environment. Or perhaps drought conditions left them little choice. Whatever the driving force, the fossil record indicates that feeding on plants wasn't even an option for the first terrestrial vertebrates.[12] By the time they arrived, plant cell walls had become encased in cellulose, an essentially indigestible organic compound.

Cellulose requires either mechanical destruction or digestive enzymes like those available to grazing animals (and termites) through symbiotic relationships with gut-dwelling microbes. Further, after eons serving as the base of the earth's aquatic food web, plants and their predecessors may have already had a jump on chemical defense, although the same might be said for the counter-defenses of their predators.

Offensive Plants

My first introduction to plant toxicants came by way of latex. This sticky white substance oozed from the wounds my father inflicted on the large rubber tree that threatened to take over the entryway to our home, each time he pruned back the branches. As he plugged the cuts with a dab of earth, he'd warn me, "Don't touch, it's toxic," which only piqued my curiosity. Latex, it turns out, contains a mixture of alkaloids, cardiac glycosides, terpenes, and other plant metabolites, depending on the species. It is a toxic brew offering protection against a broad spectrum of herbivores and pathogens. It has also been used medicinally for centuries. Consider the white latex of *Papaver somniferum*, the poppy best known for alkaloids morphine and codeine, or *Jatropha*, a latex-producing plant that derives its name from the Greek words for "doctor food."[13] Depending on the species, various chemicals in *Jatropha* latex are antimicrobial and anti-inflammatory, and may act as an anticancer drug, blocking excessive cell growth and possibly inducing apoptosis (programmed cell death) in cultured cancer cells.[14] Plant-derived drugs are very often plant toxicants; we've just learned to balance the benefits with the side effects, and even then, as we will see toward the end of this chapter, we can never be sure of our ability to maintain this equilibrium.

One of the oldest known biological toxins is saxitoxin. It is produced by cyanobacteria, which has long served as the base for many food webs. The genetic record suggests that this neurotoxin, commonly associated with paralytic shellfish poisoning and the occasional puffer fish poisoning, is more than two billion years old.[15] Saxitoxin was produced well before there were predators with neurons to be paralyzed or even neuronal sodium channels. (This might suggest that its role as a toxicant may have evolved secondarily to other physiological functions in these cyanobacteria.)[16] Saxitoxin is just one example of a toxin produced by bacteria that engage in symbiotic relationships with plants and animals, and perhaps defend their hosts in exchange for food and shelter.

Cyanobacteria may also have been important for the survival of cycads, one of the oldest known representatives of seed-producing land plants. Cycads harbor the neurotoxin beta-methylamino-l-alanine (BMAA) in their seeds and roots. This chemical has a tenuous association with neurodegenerative diseases like Parkinson's and ALS (Lou Gehrig's disease) and causes more immediate toxicity by interacting with animals' receptors for glutamate, an important neurotransmitter. The source of BMAA? While it appears to be produced by some cycad species, in others it is produced by a species of cyanobacteria.[17] The presence of BMAA in cycads raises questions about the evolutionary genetics of this chemical. Does BMAA act as an internal signal (glutamate receptors do occur in plants), is it a protective chemical, or is it both?[18] And is it possible that BMAA production in cycads capable of producing their own BMAA was derived in some way from their symbiotic bacteria? Whatever the origin and role of BMAA, we know that eventually plants came into their own, as they began producing some of the most potent toxins on Earth—many of which we now depend on as drugs.

Mining plant chemicals for potential drug activity is big business. Today, at least a quarter of all pharmaceuticals are derived directly or indirectly from plant secondary metabolites,[19] a diverse category of chemicals including many alkaloids that are stowed away in roots, shoots, seeds, and leaves. Secondary metabolites by definition are not required for growth and survival, and although their function in plants is not always clear, many of the tens of thousands of secondary metabolites are known to be poisonous. These poisonous metabolites are referred to as allelochemicals.[20] A highly varied group, allelochemicals include those mentioned earlier like the alkaloids and glycosides, in addition to quinones and tannins, and less familiar chemicals like hydrazines and saponins. These are chemicals produced, selected, and maintained presumably because they affect fitness: avoid being eaten so you can reproduce. As such, many of these defense chemicals are thought to be influenced by directional or positive selection—a process that in turn places them closer to the toxic (rather than nutritional) end of the so-called plant-chemical continuum.[21] And it is *because* these plant chemicals are bioactive—capable of activating, blocking, or otherwise interfering with normal cellular function—that we take advantage of them to rein in sickness and disease. This is particularly important these days, with concerns about rapidly declining plant biodiversity.

It is no mistake that many allelochemicals also interfere with normal cellular function, suggests the pharmaceutical biologist Michael Wink. Offering an intriguing observation about the nature of these plant chemicals, Wink writes, "Structures of these allelochemicals appear to have been shaped during evolution in such a way that they can mimic the structures of endogenous substrates, hormones, neurotransmitters or other ligands."[22] The effect of ingesting such plants can be striking. After a mysterious rash of stillbirths and infertility struck sheep in Australia in the 1940s, scientists later identified the culprit as ladino clover, a species rich in the plant estrogens genistein and coumestrol.[23] Such steroids may not only act as chemical messengers in their host plants, but also might provide population control for their predators. This dual role may be evolutionarily important. Wink proposes that "if a costly trait can serve multiple functions (and the maintenance of the biochemical machinery to produce and store secondary metabolites is energetically costly), it is more likely that it is maintained by natural selection."[24] Whether these particular plant chemicals fall into this category remains to be determined, but there are certainly benefits to controlling predatory individuals or populations while conserving energy.

More obvious than plant steroids are the defensive chemicals that act rapidly to deter predators. Some of these defenses interfere with pathways and chemicals in animal nervous systems, which of course plants do not have. This divergence of susceptibility may save a plant energy that would otherwise be required to sequester a poison to keep its own cells safe from harm. Perhaps it is no wonder that neuroactive plant chemicals abound, from the alkaloids like caffeine and tetrahydro-beta-carbolines in chocolate, to those that are potentially rewarding and therefore more highly addictive, like cocaine and nicotine. That a consumer might be rewarded by a chemical that may have evolved to deter is an interesting paradox—unless the consumer also provides benefits to the producer of that chemical, like ensuring the survival and reproduction of a host species.[25] While this is certainly true for some human-plant relations, many of these chemicals are far older than mammals and are certainly older than humans. This suggests either that relationships based on neuroactive rewards are a relatively modern phenomenon, or that we have much to learn about the brains of our tetrapod ancestors.

Finally, when it comes to plant poisons, all species are not created (nor have they evolved) equal. We know that some species are capable of co-opting a plant's poison for its own protection. A plant toxic to many insects is coveted by the monarch butterfly to make itself toxic to potential predators. Puffer fish feed on dinoflagellates loaded with tetrodotoxin and saxitoxin, which can turn a meal of the fish deadly, yet they suffer no obvious consequence. These two examples are the result of highly evolved host-predator relationships, involving the sequestration of a noxious chemical in monarchs and evolutionary changes to muscle sodium channels in puffer fish, which in each case provides immunity to their poison cargo.[26] A more domestic example of interspecies differences is the susceptibility of dogs to chocolate, or more specifically the plant alkaloid theobromine. Most dog owners know to keep the "good" chocolate (high in cocoa) away from their canine companions. Dogs are less efficient at detoxifying chocolate's theobromine than humans are; when combined with caffeine, theobromine can be fatal to them. (A recent proposal to use this combination to control coyote overpopulation gives a whole new meaning to "death by chocolate.")[27] This difference between humans and canines is in part a manifestation of differences in CYP enzymes. These kinds of variations define not only species susceptibility, but individual susceptibility as well. Some of us are more susceptible and some more resistant to toxicants—as exemplified by King Mithridates VI of Pontus.

King Mithridates (120–63 BCE) was legendary not only for orchestrating the Mithridatic Wars but also for his apparent resistance to poisoning, a common form of assassination in his day. Aside from *being* an expert poisoner, Mithridates believed that by consuming plant poisons he would become resistant to any other assassin's poison. The success of this strategy (Pontus, it is written, could not be killed by any herbal potions, and in the end died by the sword) has been attributed in large part to increased concentrations of plant-metabolizing CYPs induced by his herbal potions.[28] What we now know about certain CYPs, particularly those induced by certain herbal chemicals, adds credence to this story. Humans and other animals can respond quite differently to chemical toxicants as a result of species, diet, age, sex, and genetics. Perhaps Mithridates benefited from his herbal potion, or perhaps he was blessed with good genes—or both.

While the nature of the ancient king's resistance will forever remain a mystery, the recent advent of genomics has opened many doors to understanding interspecies and interindividual differences in CYP-mediated chemical metabolism. Until recently, our understanding of CYPs and chemical metabolism has come from the top down: how each individual or species responds to chemical challenges. This is an inadequate approach when our goal is to understand how life responds in general to toxic challenges, and when we rely on test species for predicting human responses to drugs and chemicals. And this is where evolutionary history comes in.

"If you're going to use test animals [for drugs and chemicals], you've got to understand how the metabolic machinery they have works. If you understand that system and the *evolution* of that system, the better you will understand differences between species," stresses the biologist John Stegeman. That said, the rationale for focusing on the evolutionary history goes well beyond drug and chemical testing. "It is essential," says Stegeman, "for understanding how organisms interact with their environment so that *when* there is, say, a Gulf oil spill, you know something about how life will respond."[29]

With that, we now turn to the evolution of CYPs. Given their essential role in detoxification, an evolutionary history dating back over a billion years ought not be surprising. However, what *is* surprising is the apparent role of an ancestral CYP in biochemical synthesis, rather than detoxification. And, as we shall see, many descendants of that CYP now play opposing roles—some producing and others detoxifying the very same chemical.

The CYP Defense: Animals Fight Back

Cytochrome P450s were first identified nearly fifty years ago when Ronald Estabrook, a pioneer of CYP research, characterized a key enzyme in the synthesis of 17-hydroxyprogesterone, a precursor required for the synthesis of many other hormones, from cortisol to testosterone. Little could he have imagined that decades later, the CYPs would be identified as the largest protein family known, with some seventeen thousand distinct varieties cataloged to date.[30] And while many are involved in steroid biosynthesis and other physiological processes, they are far outnumbered by CYP enzymes involved in the metabolism and transformation of toxic chemicals. That fact that all these CYPs (no matter their function) share a highly conserved region at one end of the protein suggests a single common ancestor for this amazingly prolific family.[31] Structurally, CYPs are iron-containing proteins that incorporate atmospheric oxygen into organic substrates (biochemicals) by splitting atmospheric oxygen and adding a single hydroxyl group while the remaining oxygen is reduced to water.[32] This use of atmospheric oxygen has led some to suggest that CYPs may have detoxified oxygen, creating oxygen-rich sterols as by-products.[33] Whatever the original function, when it comes to detoxification reactions mediated by CYPs today, the addition of oxygen is often just the first step in a multistage process. This complex process involves a network of other detoxification enzymes working in concert (which are discussed in greater detail in chapter 8), and rendering toxic chemicals increasingly water-soluble and therefore more amenable to excretion.

One popular candidate for this last common ancestor is CYP51—an essential enzyme intimately associated with the biosynthesis of sterols—found in all kingdoms, from bacteria to animals. Its occurrence in most eukaryotes (insects curiously have lost CYP51, and do not synthesize sterols but obtain them through their diet), and marginal occurrence in certain species of bacteria, reveals the ancient pedigree of this CYP.[34] If we imagine this early CYP's role in sterol production, it is easy to see how the descendants of this enzyme eventually became integral to the synthesis of many different sterols, from the cholesterols and sex steroid hormones in animals to phytosterols in plants and ergosterols in fungi.[35]

How does an enzyme involved in sterol synthesis evolve into one of the most important families of defensive enzymes? One scenario is that if sterols were among the first chemical deterrents (and there are plenty of toxic sterols), then co-opting an enzyme already capable of interacting with sterols to help remove them would be of some benefit. An alternative explanation invokes the role of sterols as chemical messengers that eventually required deactivation. This hypothesis is favored by the molecular toxicologist Mark E. Hahn, whose research on the aryl hydrocarbon receptor (AhR) involved in the control of certain CYPs is highlighted in the following chapter. "Once the messenger's done its job and signals the receptor," speculates Hahn, "you don't want it sticking around."[36]

Eventually, CYP and its substrates diversified beyond sterol precursors, and became involved in the synthesis *and* the breakdown of myriad chemical compounds. This provided life with a highly adaptable defense in the war between plants and animals. While not much is known yet about the CYP transition from synthesis to detoxification, there are a few key points that can be identified on the CYP family tree.

The first split occurred some six hundred million years ago, when multicellular life diverged into the deuterostomes and the protostomes. Each carried with them a shared CYP, which subsequently diversified into what is now a remarkable example of genome plasticity or flexibility, an evolutionary phenomenon driven in part by nature's propensity for both gene birth and gene death. The result is more than 1,400 CYPs in vertebrates, 2,000 in insects, nearly 3,000 in fungi, more than 1,000 in bacteria, four dozen in archaea, and a whopping 4,000 CYPs (and counting) in plants.[37] Built on the foundation of research by scientists like Ron Estabrook, the CYP field is booming. The CYP gene family is so large that the molecular biologist and CYP researcher Rene Feyereisen has suggested that it can "serve as a model for gene family evolution, perhaps one where knowledge of specific functions of P450 can bring deeper insights into the mechanism of gene family evolution."[38]

With this degree of enzyme diversity and discovery, simply naming and mapping phylogenic relationships between CYPs could become an overwhelming task for mere humans. Yet it is a task to which the geneticist David Nelson has dedicated his career, as evidenced by his "Cytochrome P450 Homepage," well worth a visit by those curious about the breadth of CYP enzymes. CYPs are classified, in general, based on the similarity of amino acid sequences. Those with greater than 40% similarity are placed in the same family (CYP1, for example); those with greater than 55% similarity belong to a subfamily (CYP1A), and within a subfamily specific genes are numbered (CYP1A1).[39] But when a single species has more than a dozen families, each with tens of genes, which may be related to those of another species, which also has dozens of families filled with different CYPs, cross-species comparisons of CYP families quickly become unwieldy. To add to the confusion, some CYPs, like siblings separated at birth, share a common ancestor but end up behaving quite differently, while others lack a common ancestor but end up appearing quite similar. And so, another way to rein in this sprawling family of CYPs is to place each *gene family* into one of four clans (clan 2, 3, 4, or mitochondrial) sharing an ancestral CYP.[40] Clans 2 and 3, for example, tend to contain many of the CYPs involved in detoxification.

The CYP Clan 2 includes one of the more intriguing detoxification enzymes, the mammalian CYP1A1—a member of the CYP1 family found only in deuterostomes, at least as far back as the sea squirt.[41] Among the first to be characterized as a "detoxification enzyme" decades ago, CYP1A1 presents an enduring puzzle. Like other in-ducible enzymes, CYP1A1 increases markedly following exposure to specific substrates: in this case, common environmental contaminants, including polyaromatic hydrocarbons (PAHs) (associated with both industrial and natural combustion products, and as such, present throughout life's evolution); PCBs; and some common plant chemi-cals. Part of the enzyme's mystery is the role of the AhR in its induc-tion. Many of the chemicals listed above have a high affinity for the AhR, and when they join with the receptor they produce increased amounts of enzyme capable of metabolizing these contaminants. But there's a catch. Particularly in the case of some PAHs, the first pass through the CYP1A1 system transforms these chemicals into potent carcinogens. It is only after a subsequent pass and further metabolism by other detoxification enzymes that these chemicals are rendered rel-atively harmless and excreted. This dual role for CYP1A1 in both ac-tivation and deactivation prompted early researchers to refer to the system as a "double-edged sword," helpful for detoxification and ex-cretion of some chemicals, some of the time.[42] Then there is the rela-tionship between CYP1A1 and the AhR. Mark Hahn and others (dis-cussed in detail in the following chapter) believe that the AhR likely originated as part of an ancient chemical sensing system, and that en-zymes like CYP1A1 may have evolved subsequently, perhaps as regu-lators of chemicals signaling the AhR.

If this book had been written in the early 1980s, when relatively few CYPs were known, and the CYP1A family's response to environ-mental contaminants dominated the research scene, CYP1A enzymes would certainly have been the focus of a chapter on the evolution of detoxification. But advances in CYP research and genomics now reveal an abundance of *other* CYPs perhaps even more important for detoxi-fication than the CYP1A enzymes. While many of these, like CYP1A, belong to Clan 2, many others belong to Clan 3. These are the CYPs that have undergone a virtual explosion in number beginning roughly four hundred million years ago, at the dawn of the plant-animal wars.[43]

Reflecting on the effects of this war on life's CYP complement, Jared Goldstone, a molecular biologist whose research focuses on the evolution and function of CYP genes, writes, "I think there has been an 'evolutionary arms race' between prey and predators, resulting in ei-ther substrate diversification of a CYP line (for example, CYP3 in mammals) or sequence multiplication and diversification. . . . There is a continuous 'birth-death' evolution of CYP diversity going on, with some endogenous functions being strictly maintained by one (or a very few) CYPs, and diversity evolving for those CYP lineages that potentially interact with environmental chemicals [but may also maintain a crucial endogenous role]."[44] Two members of Clan 3—one family unique to insects, the other a workhorse of vertebrates—provide us with a portal to observe the ongoing warfare and its consequences. This portal looks out onto old farm lots and fields, city streets and cracked pavement. We conclude this chapter with a quick glimpse.

Dining with Impunity? Furanocoumarins and CYPs

Several years ago, the local department of health raised the alarm about a highly toxic invader threatening to take over the old farm fields and roadsides of our town: giant hogweed. A noxious member of the parsley or carrot family, hogweed looks like a grotesque, overgrown version of its cousin, Queen Anne's lace. Both are members of the Umbelliferae plant family, so named for the umbrella-like flower. Hogweed, with its dinner-plate-sized leaves, eight-foot-high stalks, and two- to three-foot-wide flower heads, bleeds a sap that causes blistering of the skin and temporary or permanent blindness, should it make its way to one's eyes.

The offending chemicals are furanocoumarins, secondary plant metabolites that require one particular set of CYPs for synthesis in plants (CYP7), and another for detoxification in insects (CYP6, a member of Clan 3). Within the shoots and leaves of certain umbelliferous plants, members of the CYP7 family play an essential role in furanocoumarin production. These chemicals are *linear* in structure and become toxic upon exposure to UV light, a strategy that works well in a sunlit field.[45] Although these phototoxic furanocoumarins deter most grazing animals, some insects, including members of the swallowtail butterfly family, feed on the toxic leaves with impunity.

Perhaps in response to the emergence of resistant insects, some Umbelliferae like cow parsnip and purplestem angelica ramped up their own defense, producing a novel furanocoumarin.[46] Relying again on other CYP7 enzymes, these plants added another form of the chemical to their arsenal (this one angular rather than linear), further narrowing the pool of potential predators. This development illustrates the arms race between umbellifers and swallowtails, and is the subject of more than thirty years of investigation by the entomologist May Berenbaum.[47] Back in the early 1980s, a survey of furanocoumarin chemicals and their precursors (the hydroxycoumarins), along with butterfly larvae belonging to the *Papilio* genus, prompted Berenbaum to suggest, "It seems eminently reasonable that insects that feed on plants with hydroxycoumarins are most likely over evolutionary time to encounter plants containing furanocoumarins and thus are most likely to evolve resistance to them. Similarly, since angular furanocoumarins are not commonly produced in the absence of linear furanocoumarins, insects that feed upon linear furanocoumarins are most likely to encounter angular furanocoumarins and become resistant to them."[48]

Berenbaum hypothesized that the ever-increasing capacity of insects to tolerate these toxic chemicals might in turn effect evolutionary changes in plant chemistry.[49] But the underlying biochemical pathways leading to these changes were unknown. At that time, little was understood about the diversity of CYPs, or their role in detoxification of furanocoumarins or in the evolution of resistance. Nearly thirty years later, Berenbaum and colleagues provide insight into the evolutionary course of CYP enzymes involved in furanocoumarin detoxification. It is an evolutionary history shaped by coevolution, adaptive diversification, and gene duplication.[50] There are some key findings about the roles and responses of CYP in the black swallowtail larva (*Papilio polyxenes*). Unlike many other swallowtails, the black swallowtail seeks out and feeds exclusively on furanocoumarin-containing plants. This suggests a unique metabolic capacity. An analysis of different *Papilio* species with a range of dietary preferences reveals that CYP6B and related enzymes are important, perhaps even essential, parts of the detoxification arsenal.

Unique to insects, the CYP6 family may have evolved as a response to plant chemicals, and in some insects it is quite prolific.[51] These dramatic expansions of specific lines in the CYP6 family and others, suggests Rene Feyereisen, are best described as "blooms."[52] The CYP6 bloom in swallowtails now includes enzymes with a range of activity toward furanocoumarins, low to moderate activity in species rarely encountering the toxin, and more efficient forms (CYP6B1) in the black swallowtail.[53] CYP blooms provide raw material for an evolutionary response to new challenges, employing not only gene duplication but also diversification—an important source of new enzymes and enzyme function. And those most likely to "bloom," propose John Stegeman and Jed Goldstone, are those which appear *not* to have tightly defined physiological functions (like steroid synthesis), including certain members of CYP Clans 2 and 3.[54]

As CYPB6 enzymes became essential for the survival of exclusive herbivores like the black swallowtail, they were likely retained through the process of purifying selection, whereby deleterious alleles (those coding for a less-efficient form of CYP6B1, for example) were eliminated. In turn, this strict control of CYP6B1 may have allowed a duplicate gene in the form of CYP6B3 to evolve with less restriction as a sort of "backup" enzyme. The combination of CYP6B1 and CYP6B3 enables black swallowtails to detoxify an array of furanocoumarins more diverse than it might *normally* encounter.[55] This is a good strategy, particularly in rapidly changing environments, where Queen Anne's lace may dominate today but hogweed may take over tomorrow.

Observing the choices of these swallowtails and others in a summer field provides us a glimpse of a quiet warfare ongoing for hundreds of millions of years. This battle may equip insects with a remarkable capacity to metabolize not only plant toxins, but also industrial pesticides. Given their evolutionary history, and our current understanding of the extensive variety of CYP enzymes, it is no surprise that insects can develop pesticide resistance—or that the CYPs are considered among the most important mechanisms for pesticide resistance in insects.[56] We can only imagine how different the world's experience with pesticides and insect resistance would be had this capacity been appreciated many decades ago, as plants and insects notoriously developed resistance to pesticides like atrazine and DDT.[57]

While furanocoumarins have drawn the attention of entomologists, these chemicals have also garnered the attention of pharmacologists and physicians. We may not graze on purplestem angelica or cow parsnip, but those of us who enjoy a glass of grapefruit juice now and again ingest our own share of furanocoumarins. Most of us can do this without ill effect as long as we don't mix our grapefruit juice with any one of a few dozen drugs. These include anticonvulsants, cholesterol-reducing drugs like Lipitor, and immunosuppressants: all metabolized by one of the most important human CYPs for drug metabolism, CYP3A4. (CYP3A4 is mediated, in turn, by a receptor referred to as PXR, mentioned in the following chapter.) This is because the furanocoumarins in grapefruit juice *inhibit* CYP3A4, and as a consequence prolong the half-life of dozens of common drugs, resulting in toxic, even lethal, concentrations.[58] This reaction has taken many by surprise, including my father, whose plasma concentrations of the blood thinner warfarin (or Coumadin) became increasingly erratic after he had added several glasses of grapefruit juice to his routine (this was years ago, before the relationship between furanocoumarin and metabolism was well understood). A single glass of juice can cause inhibition in under an hour, and the effect can last for days. This is because the furanocoumarin contained in grapefruits essentially combines irreversibly to CYP3A4, necessitating synthesis of new enzyme.[59] It is such a striking and dependable reaction that some drug developers are now considering taking advantage of furanocoumarin's inhibitory power by turning it into a "drug booster." By prolonging the activity of certain drugs, they suggest, physicians may be able to reduce required dosages. This is an intriguing concept, particularly for rare or expensive drugs,[60] although one that would require detailed knowledge of an individual's CYP3A4 enzyme profile. It also reminds us of the vast number of CYP substrates, inhibitors, and inducers lurking in our foods. This is particularly important today, as consumers try to improve their health with new foods as well as herbal and pharmaceutical medications: certain combinations could be life threatening.

CYPs have clearly evolved into our detoxification workhorses. In the words of May Berenbaum, they are "the consummate environmental response genes, evolving and diversifying to a certain extent in the context of both interspecific and intraspecific interactions."[61] This sprawling system with its roots in sterol synthesis has evolved into one of the most powerful defenses we have against plant, synthetic, and industrial chemical contaminants. It is an intriguing system, quite unlike the older and perhaps more basic-to-life defensive systems discussed earlier in this book, including DNA photolyase or catalase, which have been evolutionarily conserved for billions of years. It is a system with the "genetic freedom" to evolve. As we will see in chapter 9, its potential for rapid evolution may help some species survive contaminated conditions unlike any experienced by their ancestors. Yet even robust systems like the CYPs can quickly be outmatched by novel chemicals, high concentrations of existing chemicals, or complex chemical mixtures. The effect of these evolutionary mismatches may be as inconsequential as a jittery afternoon induced by too much caffeine or, in the case of cigarette smoke, as devastating as increased susceptibility to lung cancer.

This radiation of CYPs, kicked off by the evolution of animal and plant interaction roughly 450 million years ago, continues today as plants evolve resistance to herbicides and insects to insecticides. While understanding specific CYP systems no doubt provides us with greater insight into life's response to toxic chemicals, for some CYPs the enzyme itself is only part of the story. The other half is the receptor controlling the induction or production of the CYP enzyme. In the following chapter we explore the evolution of receptors, from the AhR responsible for induction of specific CYPs to the steroid hormone receptors with no known role in detoxification, yet whose incidental response to environmental contaminants has resulted in one of the newest fields of environmental toxicology: endocrine disruption.

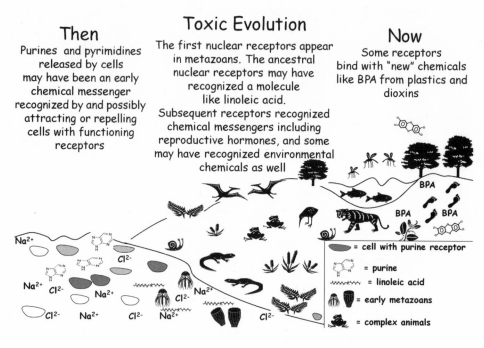

Then

Purines and pyrimidines released by cells may have been an early chemical messenger recognized by and possibly attracting or repelling cells with functioning receptors

Toxic Evolution

The first nuclear receptors appear in metazoans. The ancestral nuclear receptors may have recognized a molecule like linoleic acid.
Subsequent receptors recognized chemical messengers including reproductive hormones, and some may have recognized environmental chemicals as well

Now

Some receptors bind with "new" chemicals like BPA from plastics and dioxins

Na^{2+} Cl^{2-} Na^{2+} Cl^{2-} Na^{2+} Cl^{2-} Na^{2+} Cl^{2-} Na^{2+} Cl^{2-}

BPA BPA BPA

= cell with purine receptor

= purine

= linoleic acid

= early metazoans

= complex animals

Chemical messengers and the evolution of nuclear receptors.

Chapter 7

Sensing Chemicals

Chemical transmission, which utilizes small molecules for cell-to-cell information transfer, was an essential evolutionary step, which allowed continuous progression of life forms.

Geoff Burnstock and Alexej Verkhratsky

Over the course of billions of years, receptors and the organisms in which they function have been evolving endocrine systems that are astonishing in their complexity, diversity, and biological importance. Protecting the life-forms and ecosystems that have emerged from this evolutionary process will require that our policies take account of these characteristics.

Joseph Thornton

The challenge of surviving in an ever-changing chemical environment, and maintaining an internal environment separate from the outside, requires that life sort out the useful from the harmful chemicals. Hundreds, perhaps even thousands, of chemicals constantly enter and exit across cell membranes. Some of these chemicals become part of life, others become metabolic waste, and still others are rapidly transformed and excreted as toxic waste. Just as we are able to tune out and ignore "conventional conversations" between two people in a crowded café, but are bothered by one-sided cell phone conversations, life manages to respond to a small subset of chemicals while effectively ignoring others.[1] That living things are able to do this is an exquisite feat of chemical analysis and selectivity that requires, in part, a diversity of chemical-sensing receptors capable of distinguishing signal from noise.

Receptors that evolved in response to ancient environmental signals continue to this day to process and direct many of our responses to endogenous (internally produced) and environmental (external) chemicals. The cells in a teenage boy, bathed in a chemical soup composed of hormones, neurochemicals, and nutrients, depend on androgen (male hormone) receptors to separate the hormones from the "noise" as his voice deepens and hair sprouts on his chin. The cannabinoid receptors in the human brain respond to endocannabinoids, small lipid-soluble chemicals produced by the body to deliver messages of appetite, mood, and memory.[2] This system also responds to THC, the psychoactive plant terpenoid produced by the *Cannabis* plant—perhaps for its own defense. Both of these receptors, like many others in the body, bind not only endogenous signals or specific environmental signals, but also incidentally with chemicals in drugs, consumer products, and the environment as affected by industrial activity. Like those one-sided cell phone conversations, these are signals we cannot seem to ignore.

Our receptors are challenged with thousands of new chemicals. These unprecedented conditions present a constant challenge as receptors process and identify key signals. Understanding the evolutionary history of key receptors—the early chemical signals, their diversification over time, and the signals to which they respond today—allows us to better predict their response to today's chemical environment. In this chapter, we trace the evolutionary history of two very different receptors—the estrogen receptor, which evolved to respond to life's internal messages, and the aryl hydrocarbon receptor, which appears to have evolved responding to both internal and environmental signals. We begin by briefly considering some of the oldest receptors and signals.

Beyond Locks and Keys: Receptors and Ligands

For too many years, biological receptors have been likened to locks, opened by ligand or chemical keys. In reality, many receptors are far more complex, comprising a large group of diverse proteins, which act more like shape-shifters, capable of translating chemical information and passing the message on to other parts of the cell body. Mechanistically, when a ligand binds with a receptor, it tends to stabilize the protein in a particular shape. Or it may bind, and then be transferred from one receptor protein or carrier protein to another, influencing each—as one jigsaw puzzle morphs into another. These changes initiate chain reactions of biological responses. Some are as simple as opening a passageway across a cell membrane; others are far more complex, resulting in the turning on or off of a cascade of genes.

The most ancient chemical messengers were likely common and fairly simple. These early ligands probably included purines and pyrimidines, the common nitrogen-containing carbon rings incorporated into DNA bases, or into adenosine triphosphate (ATP)—the high-energy molecule that is life's energy currency.[3] These compounds are released both by living and dead cells. Then, as now, they likely conveyed basic information from one cell to another, whether the message was "I'm here, too, life is good," or "Danger, unfavorable conditions ahead." Today, ATP is recognized by almost all cell types across life's divisions—and may well be life's most pervasive and universal chemical signal.[4] These early chemical signals, write the neuroscientists Geoff Burnstock and Alexej Verkhratsky, required that the messenger easily move through an aquatic environment: "Choices for these diffusible messengers were only a few: they can be ions or small diffusible molecules. Ions can be excluded from extra-cellular communication pathways because of their high background concentration in the primordial seas, and thus only the relatively small soluble molecules existing in abundance within the cells can be employed."[5] In other words, life's messengers must come from life and they must stand out from the background. Although ATP's first role may have been as *the* "energy currency," at some point the molecule began interacting with cell membrane proteins and propagating a signal, an interaction that has been retained for billions of years.

The advent of receptors, including those for ATP, would have brought an unprecedented level of environmental control to cells that were otherwise at the mercy of their surroundings. While Burnstock and others continue to sort out the details of ATP receptor binding, current evidence points to at least two different types of protein receptors for ATP identified in protozoa, humans, and all life-forms in between. (ATP receptors have yet to be identified in bacteria, although some species are known to respond to purine and pyrimidine signals.) One ATP receptor family referred to as P2X, embedded within the cell membrane, opens a channel across the membrane upon binding with its ligand, much like a subway turnstile opens with the correct token. This allows sodium and calcium to flow into the cell. The other, P2Y, also sits across the cell membrane. But rather than simply opening a channel, it initiates a chain reaction leading to the release of calcium ions inside the cell.[6] Scientists have recently discovered that these receptors are critical in mediating inflammation, and at least one member of the ATP family has caught the eye of drug developers, who envision creating a more modern signal for an ancient system. There are obvious advantages to this type of drug design. But given evolutionary history and the universality of these receptors, inadvertent releases of *new* or artificial ligands for this receptor could go awry. This is of particular concern in light of the unexpected impacts of hormones and pharmaceuticals released into waterways through our urine.

While both simple and complex cells share ATP receptors, critical activities became partitioned as life evolved more complicated structures. With the advent of internal membranes, organelles, organs, and other structures, individual cells and eventually multicellular bodies needed a means of internal communication. The capacity of a body to tune in relevant signals and tune out the rest became essential for survival of the whole. With these changes in living form came proteins that could translate chemical signal into gene action by combining with DNA and activating or repressing critical genes. These nuclear transcription factors, or nuclear receptors, are critical for inter- and intracellular communication, maintaining homeostasis, and responding to external cues. They may well be the only receptors that connect chemical signals from the metabolic environment to gene regulation.[7] Including estrogen receptors (ER), androgen receptors (AR), growth hormone receptors, receptors that control CYP expression, and others, nuclear receptors are incredibly sensitive, distinguishing signal from noise even when the signals are in the parts per trillion or less. Yet these receptors also provide us with some of the most striking examples of mistaken chemical identity. Some inadvertently respond to small amounts of "new" chemicals, including pesticides, plasticizers, and other commercial products.

Turning on DNA: Nuclear Receptors

Nuclear receptors like ERs and the PXR (which controls expression of CYP3A enzymes) set in motion specific biological responses. And although the receptors tend to be highly conserved, sometimes the outcome of activation may differ dramatically between species or individuals within a species, depending on age or sex. Estrogen, for example, combines with the estrogen receptor to induce egg yolk production in fish and puberty in adolescent girls. Chemicals that combine with the AhR induce certain CYPs in some tissues in some vertebrate species but not others. Sometimes these differences make sense. Egg yolk production in response to estrogen can be reconciled, to some extent, with maturation in human females—both actions are necessary for successful reproduction. Other times these differences seem more scattershot. One species may respond to a drug or chemical by producing a specific CYP, whereas another will not. Though two species may share a common ancestor and hence a common ancestral receptor or enzyme, once they part ways on the family tree, the branches evolve independently.

The receptors observed in living things today are the result of ongoing evolutionary processes, including gene conservation and gene duplication. Some genes and biochemical processes are highly conserved, like those associated with reproduction and maturation, while others have greater freedom to diversify, and still others are lost. Conservation of genes explains why estrogen receptors in both striped bass and the women who fish for them respond to estradiol (a potent form of estrogen). Duplication, in contrast, is one source of raw materials for evolutionary change, providing an opportunity for diversification. When it comes to receptors, this phenomenon has provided us with forty-eight different nuclear receptors, which respond to at least as many different chemicals.[8] This is in contrast to the two nuclear receptors (at most) identified in sponges and ctenophores—the iridescent "comb jellyfish" we might see on a summer's day pulsing along in the bay—while nematodes clock in with well over two hundred members of their nuclear receptor family.[9] Nematodes aside, vertebrates have a large number and diversity of receptors, which are thought to have derived from two different incidences of whole genome duplication (or serial genome duplication).[10] Such large-scale duplications would have affected almost every defensive system discussed so far, contributing to the evolution of new and sometimes unique defensive attributes.

Nuclear receptors, unlike the chemical defenses discussed in the previous section, appear to be an exclusively metazoan innovation, stemming from a single ancestral receptor (AncNR). Yet the composition and role of that ancient receptor remains a mystery. Receptors in species surviving today all share a highly conserved DNA-binding region (and a less conserved ligand-binding region), and likely shared with a distant ancestor the amino acid sequences that make up these key regions.[11] Generally, when discussion turns to receptors, it is focused on receptors with known ligands. Yet receptors referred to as "orphan receptors," with no known ligand, may provide some insight into the earliest nuclear receptors. Stuck in the "on" position, these receptors are sometimes referred to as constitutive because they continuously promote gene transcription. The behavior of these orphans has led to different ideas about the modus operandi of the ancestral nuclear receptor and its subsequent evolutionary history. Could the superfamily of nuclear receptors, now characterized by their ligand specificity, have evolved from orphan receptors—with ligand affinity developing subsequently and independently over and over again throughout evolution?[12] Or is the ability to bind with a specific ligand not only ancestral, but the result of evolution by "molecular tinkering," whereby slight changes in the ancestral receptor's ligand-binding "pocket" result in sensitivity to a large variety of substances?[13] Perhaps even more intriguing (given our focus on chemical defense) is a recent proposal that all NRs evolved from the need for early life to respond to nutrients—chemicals that may either fulfill a cell's dietary requirements or, if in excess, become toxic.[14] This suggests these early receptors may have responded to a variety of lipophilic chemicals (fatty acids, steroids, and other small lipids) likely present in vanishingly small amounts, yet which eventually became essential to metazoan life. Given the lipophilic nature and relatively small size of many environmental contaminants, if this scenario is true, it may help explain why such a large number of chemical contaminants interact with biological receptors.

Whatever the origin of nuclear receptors, most agree that the AncNR emerged at the dawn of metazoan life and evolved into a superfamily of highly sensitive receptors. These receptors are now represented by six different families, including receptors that bind with thyroid hormones and retinoic acid, and which are involved with cholesterol (and cholesterol derivative) and xenobiotic metabolism; receptors that bind with fatty acids and retinoids; estrogen and androgen receptors; receptors involved in embryogenesis; and those with no known ligands ("orphans") involved in immunity and development.[15] All six families emerged at least five hundred million years ago, and many (but not all) emerged just before the split between the protostomes and deuterostomes.[16] As a result, all these receptors now occur across animal species, from houseflies to humans, while representative receptors from only one or two families occur in their pre-bilateran cousins (those animals without bilateral symmetry), a branch that diverged earlier. Of the six families, one of the most intriguing includes the reproductive receptors that belong to nuclear receptor sub-family 3, the estrogen (and estrogen-like) receptors.

Let's Talk about Sex

A little over twenty years ago the phrase "endocrine disruptor" was virtually unknown. Though laboratories around the country were investigating all sorts of endocrine-disrupting chemicals at the time (DES in humans was a well-known example of a potent estrogen, and raptors were making a comeback following the U.S. ban of DDT), there were few coordinated efforts. That was, however, until the zoologist Theo Colborn began knitting together results from disparate research efforts on fish, birds, and mammals.[17] The most striking (yet not all that surprising) finding of her synthesis was the commonality of effects caused by certain industrial chemicals across species. That male fish began producing egg yolk protein because they were exposed to environmental estrogens became a cause for concern not only for fish populations but for humans as well, as several investigators published studies suggesting increased reproductive and developmental abnormalities in boys and men associated with estrogenic compounds.

Two scientists whose work attracted Colborn's attention were the reproductive toxicologists Earl Gray and Bill Kelce. Their analysis on the pesticide vinclozolin highlighted the impact of chemicals that interfered with the androgen receptor (AR). Following up on an earlier industry study, Gray found that litters born to female rats exposed to vinclozolin, oddly, *appeared* to be exclusively female. In reality, the sex ratio was unchanged, but the external or secondary male characteristics of those born to the most highly exposed females had been dramatically altered. This observation eventually led to a series of studies revealing the perversity of endocrine disruption.[18] Basically, for a genetically male mammal to come out looking and functioning male, he requires in utero exposure to hormones like testosterone and its more potent derivative, dihydrotestosterone, along with a functioning AR. An embryo lacking either hormones or a properly functioning AR (or exposed to chemicals that disrupt either receptor or hormone production) will take on a female appearance, despite possessing a Y chromosome. In humans, one outcome of a non-functional or partially functional androgen receptor is androgen insensitivity syndrome, a condition resulting in the presence of female sex characteristics and absence of male characteristics to varying degrees, although sex-appropriate internal organs are retained.[19]

Vinclozolin, as it turned out, caused something like this in rats. In an elegant series of follow-up studies, both Kelce and Gray showed that breakdown products of the pesticide not only bound with the AR, but rather than activating the receptor, caused inhibition. The result was the birth of genetic males whose outward appearance was female.[20] Additional work by Kelce, Gray, and others revealed that a metabolite of the pesticide DDT was an even more potent inhibitor of the AR than was vinclozolin. Given the ubiquity of DDT and its metabolites, this was a potentially explosive finding.[21] Many of the studies included in Colborn's synthesis tended to focus on chemicals interfering with the estrogens and the ER; Gray and Kelce's work highlighted the importance of chemicals that interfered with the workings of male hormones.

Theo Colborn's efforts underscored the pervasive impact of industrial contaminants on sex steroid hormones and other hormones, including the thyroid hormone. They also revealed the lack of coordinated testing of these chemicals for subtle reproductive effects and interactions with steroid hormone receptors like the ER and AR. Her work resulted in the birth of a new field of research focused primarily on the impact of industrial chemicals on these highly conserved chemical communication systems.[22] These chemicals have the potential to affect the reproductive capacity, and therefore fitness, of a large swath of life on Earth, humans included.

Vertebrates possess a number of steroid hormone receptors that belong to a common subfamily of NRs (NR3). In addition to the AR, these include two or more different ERs, receptors for progestagens (which help maintain pregnancy), glucocorticoids (important for immune function and maintaining a variety of metabolic functions), and mineralocorticoids (which regulate mineral balance in the body). Together these receptors moderate our behavior, our sexual development and reproduction, our response to stress response, and, to some extent, our immune response. By knowing something about the evolution of these receptors and their ligands, we might be better able to predict and perhaps avoid production and release of the most egregious endocrine-disrupting chemicals.

As with nuclear receptors in general, nuclear steroid hormone receptors are thought to have evolved from a single ancestral steroid receptor. Although steroid hormone activity in mollusks and in annelid worms (both protostomes) suggests that a receptor was present in the last common ancestor before the protostome-deuterostome split, steroid hormone receptor absence from other protostomes (including insects) most likely indicates gene loss.[23] So while estrogen and androgen receptors are essential for the birds, they apparently are not for the bees. Not only was it retained in the vertebrate lineage, but the ancestral gene multiplied and diverged. This provided us with dozens of hormone receptors capable of distinguishing one hormone coursing through our veins from another, while guiding our development, growth, and behavior.

As to the function of this ancestral steroid receptor, the evolutionary endocrinologist Joseph Thornton suggests that the similarity between the sole steroid receptor identified in mollusks, annelid worms, and the ER in vertebrates points to an estrogen-like receptor. "The capacity to bind estrogens and estrogen response elements [on DNA]," write Thornton and Eick, "is clearly as old as the common ancestor of protostomes and deuterostomes."[24] So what came first, the receptor or the ligand? Is estrogen the grande dame of steroid hormones, or did another chemical messenger serve the purpose? Thornton suggests that if estrogens did evolve subsequent to the receptor, the ability of the receptor to accept estrogen as a ligand would represent a "promiscuous side activity," which industrial contaminants inadvertently exploit.[25] If not interacting with estrogen, what purpose would an estrogen-less receptor serve? Perhaps its role was simply as a sensor for signals from the surrounding environment, or from other organisms, translating and then sending them on to the animal's internal environment.[26] Though still a mystery, the accumulating body of literature on estrogen and other steroid hormone receptors points to a set of receptors that tend to be far more promiscuous than previously believed, a quality that possibly harks back to their origins.

If the original receptor was an estrogen receptor, the evolution of *other* steroidal receptors, from the glucocorticoid to androgen receptors, then required only small changes in ligand binding (mutations in just a few amino acids, for example) in order to accommodate other steroid hormones. The steroidal ligands for these receptors are all related to one another as by-products of the biochemical pathway progressing from cholesterol to estrogen (and many steps along the way are notably catalyzed by CYPs).[27] That the final product in a long chain of reactions was the first to have its own receptor might be surprising, but is not unprecedented. Thornton raises the possibility of "molecular exploitation." In this process, older molecules are recruited for new functions, followed by refinement of the receptor through directional selective pressures "driv[ing] receptors towards greater specificity for their primary ligands."[28] This diversification of receptors may have been aided through gene duplication, an event that would have provided the opportunity for small changes in receptor specificity.[29] The susceptibility to industrial and synthetic chemicals of what seem to be finely tuned receptors reflects perhaps not a "weakness" in receptor evolution, but rather the novelty of these chemicals. Thornton posits that the chemicals "fit by chance into NR binding pockets, which have *not* been selected over the long term to *exclude* binding of these substances."[30]

We now know that endocrine-disrupting chemicals are ubiquitous. There is a constantly increasing body of research that points to ever more consumer and industrial chemicals that interact with a steroid hormone receptor, from the nonylphenols that activate the ER to the antiandrogenic metabolites of vinclozolin. It is fair to say that for every steroid hormone receptor, there is an imposter: an endocrine-disrupting chemical capable of either activating or blocking its normal function.[31] Curiously, of those disrupters identified, the list of estrogenic chemicals appears to be never-ending, while few chemicals have yet to be identified as androgenic. Why? Could this be as simple as research and reporting bias focused on estrogens? Or is there something about the structure of industrial and consumer-product chemicals that increases their tendency to behave more like estrogen?[32] Or is there some fundamental difference between these receptors, perhaps a difference in promiscuity or some other remnant of their evolutionary history? "It is clear that the xenobiotics that bind ER are primarily agonists [activators] and those that bind AR are mostly antagonists [inhibitors, except drugs and steroid-like molecules]," writes L. Earl Gray about the discrepancy. "One can speculate why this is the case. Both receptors are promiscuous, but the ER appears to be more so. It is clear that this is not a bias in testing, and it is an interesting observation."[33] The biochemist Jed Goldstone suggests that estrogen, by virtue of its ringed aromatic structure, is more resistant to breakdown and chemically very different from many other internal chemicals. Evolutionarily, posits Goldstone, it may have made some "sense" for a receptor to bind with a chemical that was different rather than common. (This harks back to the argument for purine and pyrimidines as early messengers rather than commonplace ions for early life.)[34] Since the more stable environmental contaminants also possess aromatic rings, they may more likely interact with the ER rather than the AR, as testosterone does not have an aromatic ring. That industrial chemicals can bind with steroid hormone receptors is no trivial matter. Impacts at the population level have already occurred in wildlife because of these evolutionary mismatches. Whether humans will experience the same is unknown. Yet as Colborn has shown, there is little reason to suspect we would respond differently than many wildlife species.

The development of steroid hormone receptors provides an opportunity to marvel over the intricate nature of evolution, which has resulted in ligands and receptors that exert powerful control over our bodies, actions, and thoughts. At the same time, we realize the imperfections in these systems. Diverse species are susceptible to industrial chemicals that may inadvertently combine with critical receptors. And the consequences, whether production of egg yolk in male fish or the subtly altered secondary sexual characteristics in a young boy, are only now playing out, as those of us born to this industrial world mature and reproduce. Yet with increased awareness, we may be able to protect future generations from the most egregious endocrine-disrupting chemicals.

This ability of many natural or industrial chemical contaminants to combine with steroid hormone receptors may be an accident of nature, but what of the receptors that seem to have evolved to bind specifically with potentially toxic chemicals? In addition to receptors like the AR and ER, there are receptors to a broad assortment of chemicals for which life apparently has little use, collectively referred to as xenobiotic-activated receptors. According to the toxicologist Qing Ma, these receptors share certain characteristics, including their response to foreign and potentially toxic chemicals; promiscuity; conserved structures; rapid activation and deactivation; ability to interact with a common set of proteins (involved in transport and regulation of the receptor); and ability to turn on a broad range of genes, some of which are common to different receptors.[35] Did xenobiotic-binding receptors evolve in response to toxic chemicals? Or were they co-opted from receptors dedicated to other functions? While many of these receptors have only been identified as such in the past two decades or so, one xenobiotic-binding receptor, the AhR, has been both intriguing and baffling scientists since the 1970s.

An Inexplicable Affinity

In graduate school I studied CYP1A enzymes. This enzyme family is inducible upon exposure to dozens of chemicals, including benzo[a]pyrene (B[a]P), a combustion product common to forest fires, auto exhaust, and cigarette smoke (referred to generally as a PAH); some PCBs; and dioxins. Induction (increased protein production) required interaction with the AhR. While the AhR bound with PAHs and certain PCBs, it had an oddly high affinity for dioxin. That many of the AhR's chemical ligands from B[a]P to dioxin tended to be similarly shaped, with a flat or planar chemical structure and of a similar size, made some sense. This predictable relationship between a ligand's structure, affinity for the receptor, and ability to induce CYP1A proteins (which in turn metabolized some, but not all, inducers, as discussed below) led to the development of a hierarchy of structure-activity relationships (SARs)—all normalized to the affinity of the receptor for dioxin. In time, these SARs became the basis for predicting the toxicity of certain classes of common environmental contaminants, and a boon to toxicologists faced with predicting the toxicity of sites contaminated with PCB or dioxin mixtures. Yet for those thinking about the evolution of the CYP and AhR systems, the relationship between chemical inducers and CYP1A enzymes did not always add up.

Given the potential for an enzyme like CYP1A1 to detoxify B[a]P while at the same time producing potent carcinogens along the way (the double-edged sword discussed in the last chapter), was there truly an evolutionary advantage to such a receptor, *if* detoxification was its primary role?[36] Perhaps for rapidly maturing organisms with short life spans, cancer didn't provide a powerful selection pressure. Yet plenty of longer-lived species maintained both the receptor and the CYP1A1 enzyme. And what were toxicologists to make of its high affinity for dioxin—a chemical most closely associated with industry rather than nature—and the fact that dioxin is relatively resistant to CYP1A metabolism (one reason why it so readily accumulates in many animals)? More than twenty years have passed since I was a student, and while the quest to understand the relationship between dioxin and the AhR and CYPs has led to great advances in our understanding of receptor function and enzyme induction, many outstanding questions remain. Reflecting on his experiences deciphering the AhR, the biochemical toxicologist Allan Okey, who has contributed a great deal to our current understanding of this receptor, wrote, "Our quest for the mediator of dioxin's biochemical and toxic effects has let us glimpse many important and fundamental AHR attributes through gaps in the fog. But, as is perpetual in science, beyond the shore lies *terra incognita*."[37] While we may not yet have reached the shore, as Okey observed, the journey has revealed quite a bit about this odd but important receptor.

One scientist to help clear the fog is the biochemist Mark E. Hahn. His work has illuminated some of the basic protein and ligand interactions involved in AhR function.[38] Like steroid hormone receptors, AhR is classified as a nuclear transcription factor. It also belongs to a family of receptors referred to as basic helix-loop-helix/Per-Arnt-Sim, characterized by specific folds and structural domains in the AhR protein. Once a ligand binds, like other nuclear receptors, the AhR proceeds through a series of events that can include binding and separating from other proteins. Some of these proteins chaperone the receptor into the nucleus, while others tend to repress the receptor's activity. It is a complex system. The evolution of both chaperones and repressors might suggest that just because a receptor-mediated pathway *can be* initiated by a ligand, this may not always be in the best interest of the host cell. So selection pressures have provided a few checkpoints or obstacles to help rein in activity—maybe.

Hahn and colleagues have also contributed a great deal to our understanding of the evolutionary origins of the AhR and associated proteins. According to Hahn and others, some form of the AhR likely existed in the last common ancestor to deuterostomes and protostomes. Yet, interestingly, its ability to bind with PAHs and dioxins—a feature so well characterized in fish, birds, and humans—is lacking in fruit flies and mollusks.[39] While the function of this receptor in mollusks remains unclear, it appears necessary for development of the antennae, neurons, and photoreceptors in fruit flies, confirming its origins as a chemosensory and developmental receptor. And through five hundred million years of vertebrate evolution, even as gene duplication events endowed some tetrapods (bony fish and some bird species) with multiple AhR receptors, it seems that at least some of these basic functions have been retained.[40] Aside from binding dioxin and the like, the AhR is known to regulate very basic biologic functions including growth regulation and fetal development.[41]

The fact that the vertebrate AhR also gained the ability to respond to specific ligands like dioxin and B[a]P, and induce CYP enzymes, means that all of us contain within our cells a receptor that appears to have a dual nature—which may not always serve us well. "Thus, the physiological functions of the AHR may be ancestral to the adaptive function, although new physiological functions appear to have evolved as well in the vertebrate lineage," writes Hahn. "One might conclude that the emergence of the adaptive function of the AHR is responsible for dioxin toxicity, because it led to the ability of this protein to bind PHAHs and PAHs, which could then interfere with its physiological function."[42]

The AhR's role in normal development and in the metabolism of xenobiotics raises plenty of questions about the receptor and its evolutionary history, beginning with the nature of AhR ligands (most recently, several physiological ligands have been identified) and the necessity of CYP1A enzymes.[43] Given the role of CYP enzymes in the metabolic breakdown and the eventual excretion of so many potentially toxic chemicals of both internal and environmental origin, it seems likely these enzymes became and remain important components of a general chemical defense system. And, if that is the case, then the AhR is also a component of this system, regardless of whether it started out that way.

Another fascinating feature of the AhR is the propensity for striking differences in receptor-ligand affinity—particularly when it comes to binding dioxin—between and within a species. Best exemplified by different strains of mice, some species are relatively "nonresponsive" to dioxin, while others are dozens of times more sensitive. Similarly, guinea pigs are exquisitely sensitive to dioxin, particularly in comparison with hamsters. Humans appear to fall somewhere toward the less-sensitive end of the spectrum with respect to dioxin binding (but not necessarily receptor response to other ligands).[44] These observed differences in susceptibilities appear, in part, to be a consequence of small differences in the receptors' amino acid composition. Differences in just one or two specific amino acids in the binding site can be the difference between heightened sensitivity and relative resistance.[45] Selective pressures leading to such differences for the most part remain a mystery. But there is growing evidence that at least in some species, the AhR is susceptible to rapid evolutionary changes in response to local contamination (discussed in detail in chapter 9) and, perhaps, differing endogenous ligands.[46] The fog surrounding the AhR is thinning.

We have touched on two receptors involved in chemical sensing and response to naturally occurring and industrial chemicals. These receptors raise many questions about the evolutionary paths of chemical receptor systems, their role in our lives today, and our ability to predict not only toxicity but also differences in sensitivity and susceptibility within and between species. Now consider the larger collection of receptors: other nuclear receptors or xenobiotic-activated receptors (XARs)[47] responding to internal and external environmental signals. This chapter could easily have focused on any one of these other receptors. Not only that, but many receptors, and the specific proteins with which they interact, influence one another *and* various other proteins and enzymes discussed in previous chapters. These interactions constitute a highly complex networked response to chemical challenges, as we shall see in the next chapter.

Then

Constant exposure to environmental stresses, some of which may be toxic, has resulted in a network of environmental response genes

Toxic Evolution

Over time, this network increased in complexity as environmental stressors increased and included more potentially toxic chemicals

Now

Chemicals released through industrial use and in consumer products are constantly challenging this defense network

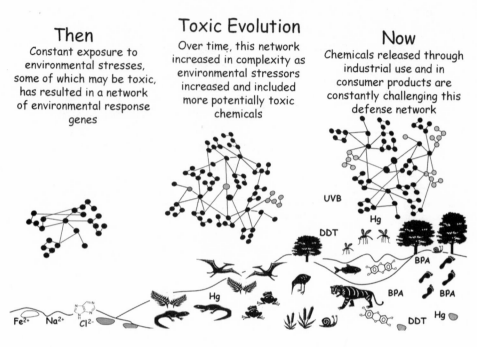

The evolution of networked defenses to environmental stress.

Chapter 8

Coordinated Defense

An important question in biology is how cells and organisms maintain homeostasis in a variable environment. The need to deal with physical, chemical, and biological stressors has driven the evolution of an array of gene families and pathways . . . that afford protection from challenges.

Jared Goldstone

The most adaptable and successful organisms, though wildly diverse in appearance and behavior, are all organized in a similar manner. Universally, they avoid the trap of centralized, top-down control by giving wide ranging power to multiple independent sensors to observe and respond to environmental change and threats.

Raphael Sagarin

The extent and content of metazoan gene repertoires are governed principally by the evolutionary turmoil of environmental genes.

Chris Ponting

A body at rest never really is. If we could envision the activity within our cells, we might liken it to a beehive or ant colony. Chemicals come in, chemicals go out. They are transported and transformed by dozens, perhaps hundreds, of proteins. Some of these proteins act as enzymes, some as chemical transporters, while others turn genes on and off as they respond to chemical signals. Some of the oxygen permeating my cells will become highly reactive, touching off a cascade of defensive proteins, produced by a genetic blueprint that first came into being billions of years ago. The lunch I ate at noon will break down into its molecular components—fats, sugars, and proteins, in addition to various elements and minerals. Some will pass through while others will be absorbed by specialized gut cells and distributed throughout my body, and still others, perhaps plant toxins, will be transformed by a series of enzymes. Oxygen will no doubt interact with some of those nutrients, and with other molecules and metabolites. The iron from last night's kale will be extracted, and perhaps bound within a heme group, which will use the oxygen to metabolize those molecules presenting a threat to my body. All these chemicals travel highly evolved pathways.

Some of these pathways have evolved over billions of years, others over a mere few hundred million, and still others more recently. Yet all have been integrated through the processes of evolution. The human body is simply one representation of a constantly changing, complex network. But to what extent can this network handle today's chemical universe? This new world includes plastics, chlorinated organics, pharmaceuticals, and engineered nano-sized chemicals, and continues to expand at a pace that outstrips anything life has experienced in the past. To adequately consider the impact of contemporary chemicals (or, in fact, *any* chemicals, the topic of the last two chapters), we must first consider the networked nature of life's chemical defense system.

Evolution weaves a complex web, and focusing on individual systems as I have throughout this book (e.g., AhR, CYP, ROS) conveys a false impression of simplicity. If we consider the evolutionary history of any single system (including those discussed earlier in this book), the genes involved are likely the product of a convoluted series of evolutionary twists and turns. Genes for one system may have emerged and diverged from genes evolved to serve different functions. In the process, perhaps these systems retained some overlap; when one system turns on, so too might parts of another, depending on the type and level of interaction. Some will share so-called chaperone proteins. Other systems and genes will share regulatory regions or DNA sequences. Although traditional toxicology has focused on the effects of single chemicals on isolated systems (phthalate plasticizers on estrogen receptors, or cadmium induction of metallothionein), the reality is that these systems are interconnected. No system works in isolation.

Further, cells are exposed to a multitude of chemicals simultaneously—be they endogenous signals, environmental signals, nutrients, or toxicants.

The single chemical, single end point system on which much of toxicology has rested is a false assumption, born of necessity, limited by available methodology, and driven by the need for numerical standards. While these approaches have served us well, life's reality is far more complicated. Rather than coping with one chemical at a time, living things deal with myriad chemicals, some toxic, others essential, and many both toxic and essential. There are countless biochemical reactions occurring within a cell at any given time. Some are coordinated while others function in opposition, scavenging for shared components or available energy. Together, these actions and reactions maintain internal stability. This balance, or homeostasis, is a preferred condition—it is common to all cells, and the chemical defense network has evolved in large part to prevent toxic chemicals from upsetting this balance.

This chapter is different from the preceding chapters, which followed the evolution of a system from beginning to end. Here we consider the larger context within which the defense systems discussed throughout this book exist: the network of environmental response genes. Additionally, I mention epigenetics, hormesis, and attractor states as examples of processes that may influence the interaction between chemicals and environmental response genes. Much of the work on these topics as they apply to toxicology is fairly recent, so an in-depth understanding of their role in the response and evolution of response genes remains to be discovered. Yet I include them here because not only are they intriguing concepts, but they also may someday prove fundamentally important. But first we need the context.

The Environmental Net

Recent advances in genomics have provided us with humbling revelations about the relationship between DNA, genes, evolution, biological complexity, and how life responds to its environment. For example, while the human gene complement was once thought to number in the hundreds of thousands, we now know that it may be closer to twenty thousand—not much different from that of a nematode worm.[1] Single-celled yeasts, in contrast, rely on six thousand genes, while coliform bacteria are defined by a little over four thousand. There are clearly many mysteries to be solved as geneticists reconcile genome similarities and differences between bacteria, yeasts, nematodes, and humans. Our concern is with those genes and gene networks that respond to both internal and external environmental pressures. These genes, whether in yeast or humans, are referred to collectively as environmental response genes.

One of the first species to have its environmental stress response (ESR) genome analyzed in any detail was the yeast, as presented by Audrey Gasch and colleagues. Gasch suggests "that a substantial fraction of each of the responses *is not specific* to the stimulus but instead represents a common response to all of the conditions tested."[2] This finding implies that, at least in yeast, there is a fundamental response to disturbances caused by environment stressors. That said, Gasch notes that there are differences in the overall response, depending on the conditions: "Each genomic expression program triggered by environmental change is unique to the specific features of the new conditions in terms of the genes affected and the magnitude and choreography of their expression, indicating that the cell precisely responds to the distinctive challenges of each new environment. Nonetheless, the bulk of each genomic expression program is accounted for by the genes in the ESR."[3] Remarkably, the response involves a large proportion of the yeast genome, roughly 14%. This means, at least in yeasts, when faced with environmental challenges such as chemical stressors, a significant portion of the genome is involved in the maintenance of homeostasis.

Unlike metazoans, Gasch's yeast lack the luxury of multiple cell types, compartmentalization, and the protective layers (both physical and physiological) that many of our cells enjoy. Like most single-celled species, yeast must maintain homeostasis despite their vulnerability to both subtle and swift environmental challenges. Yeast survive by turning on and off many hundreds of genes and by responding in a remarkably coordinated and "stereotypical" manner. One cohort of genes turns on as another is repressed, providing almost a mirror image of expression and repression.[4] This conservation and reallocation of resources to where they are needed most is a response that, most likely, applies to all cells—whether eukaryotic or prokaryotic. While some genes observed by Gasch and colleagues require specific chemical signals to touch off a response, others are induced no matter the stress. These more general responders include some of the genes and proteins discussed early on in this book: enzymes that mop up reactive oxygen species, and genes coding for proteins involved in protein folding and in the repair of DNA. These responses, suggest Gasch and colleagues, "represent physiological systems that must be protected under any circumstance." Additionally, Gasch and others have observed that stress response gene networks also provide cross-protection: exposure to one stress protects against another, unrelated stress.[5] Cross-protection is an advantage of networked responses that has yet to be fully explored. Pending further research, it might help explain the rapid development of resistance in bacteria, or why some individuals are more susceptible to specific chemicals in comparison to others, or why some chemicals appear to be more or less toxic than expected based on toxicity testing data. And, if crossover effects can be protective, could they also be detrimental? Might they influence the response network so that it is more vulnerable, rather than less? This outcome would make predictions of multiple chemical exposures difficult based on current toxicity testing practices.

How do these observations of large, and somewhat generalized, responses to stressors in yeasts translate to metazoan cells, or to humans? Because metazoan cells need not fend for themselves, should we expect a more "specialized" response depending on cell type—whether a skin cell, a liver cell, or a gut cell? At least one study using cultured human cells suggests a far more muted stress response in comparison to our eukaryotic cousins. "The absence of a strong general stress response in humans," write the geneticist John Murray and colleagues, "is in contrast with free living yeasts and bacteria which do have large common stress responses. Several key differences between these distant species might play a role in this divergence. Single cells in multicellular organisms experience different selective pressures than unicellular organisms; human cells may have a more controlled response because they may decide to undergo apoptosis [not an option for unicellular life]. . . . Moreover, human cells probably experience a less variable environment than free-living organisms, which also could explain the diminished general stress response."[6] There is also another explanation set forth by the authors, which relates to the artificial conditions of cell culture that may chronically stress the cells, thereby reducing their capacity to respond to additional stress. The differences between cultured cells and whole animals is a common caveat of in vitro research, and one that may be solved in part through cell culture advances that better mimic tissue and organ architecture, and that allow for signaling between cells and the "normal" cellular environment.[7] In other words, the biochemical and cellular networks surrounding a cell are important for maintaining normal cell function. These stress response genes, like those discussed above, are of particular interest to toxicologists. The advent of omics and associated fields (particularly toxicogenomics, as discussed in the following chapter) is providing toxicologists, geneticists, molecular biologists, and others with unprecedented opportunities to observe the turning on and off of many thousands of genes in response to environmental stressors, including potentially toxic chemicals. Interpreting the morass of data that is sure to follow, however, demands (in addition to complex mathematical and computer modeling programs) refined knowledge of stress response gene networks and their role in maintaining cell, organ, and organism health.[8] One approach to reining in genome responses may be to recognize broad categories of response genes. Genes that respond to toxic chemicals in particular, referred to as the defensome, may offer an appropriate starting point.

The Defensome

The chemical defensome is described by the molecular toxicologist Jared Goldstone and colleagues as "an integrated network of genes and pathways that allow an organism to not only mount an orchestrated defense against toxic chemicals but also maintain homeostasis— or internal stability."[9] Whether sea urchin, sea squirt, or human, these genes constitute upward of 2%–3% of the whole genome (this includes immune defense genes, which, for the sake of simplicity and sanity, I have not covered in this book).[10] The defensome encodes proteins ranging from those acting as antioxidants, oxidative enzymes, receptors, and signal transducers (as discussed in preceding chapters) to others, such as the efflux transporters, which pump potentially harmful chemicals across cell membranes and out of the cell, and which are among the most ancient and perhaps simplest defensive features on Earth. In general, genes encoding many of these responses, writes the entomologist May Berenbaum, may be characterized by their "(1) very high diversity, (2) proliferation by duplication events, (3) rapid rates of evolution, (4) occurrence in gene clusters, and (5) tissue- or temporal-specific expression. All these characteristics are consistent with responses to evolutionary pressure emanating from a highly changeable external environment."[11] We have seen all these characteristics in the systems discussed in preceding chapters. We have also seen that the external environment is not the only source of harmful chemicals. And so, in addition to defusing external chemical threats, defensome gene products neutralize or eliminate chemicals emanating from within—whether endogenous signaling molecules, like estrogens that have served their purpose, or ROS created as a by-product of CYP metabolism.

The integrated nature of the defensome suggests that exposure to one toxicant, whether concurrent or sequential, is likely to affect the response to another. As inferred earlier in this chapter, one casualty of toxicology's focus on individual chemicals is the inability to predict the effects of multiple chemicals.[12] Yet multiple chemical exposures are the norm. We are constantly exposed to a veritable chemical smorgasbord, whether through our foods, drugs, consumer products, air, or water, and we must begin incorporating networked responses into chemical testing routines.

Goldstone and colleagues have also contributed a great deal toward deciphering the defensome evolutionary history by characterizing the defensome and seeking commonalities and differences across species. We have seen that some systems are highly conserved (DNA photolyase, for example), while others diverge through duplication or are lost. Yet, even as individual *genes* may evolve, the gene *families* making up the defensome have tended toward conservation throughout metazoan lineages, whether sea anemone, zebra fish, or human.[13] This basic defensome plan, which includes all systems discussed in this book, has been retained for some five hundred million years and is now shared across deuterostomes (from vertebrates to echinoderms and a few marine wormlike species).[14]

Scientists like Goldstone strive to glimpse the past reaching back hundreds of millions of years, yet others seek out much smaller timescales: ontological, or developmental, time. Metazoan reproduction and development follows an extraordinary and highly choreographed progression from single cell to embryo to complex animal. This process refers back to the days of unicellular life, as a great majority of these cells, particularly in aquatic organisms, are released to the environment as single, relatively unprotected egg cells. Even those surrounded by multiple membranes or hardened yet permeable eggshells are far more exposed than are the cells of their adult forms. How protective is the metazoan defensome at these vulnerable stages? Is gene expression altered in contrast to their adult forms? Are a greater proportion of environmental response genes available in the egg and embryo in comparison to a fully matured and differentiated cell? Or is the production and release of hundreds, thousands, or even millions of eggs a way of hedging the bet—in the hopes that one or two offspring survive the environmental gauntlet?

Defense Networks across the Ages

Throughout this book, many of the studies I have cited refer to effects and responses in mature, adult cells. Yet just as focusing on single genes rather than gene networks is an oversimplification, so too is limiting ourselves to the defensive capacity of any one stage of metazoan development. We know that highly differentiated cell types express unique combinations of defensive genes, such as CYPs in the liver or efflux transporter proteins in the digestive tract. So how will an egg, a single cell, or an embryo encased in nothing more than a few semipermeable membranes fare as it faces environmental hazards, from UVB to petroleum products to dramatic shifts in temperatures?[15] Surprisingly, these budding life-forms appear to have highly effective defensive systems, according to the developmental biologists David Epel and Amro Hamdoun: "Equally striking are examples of how real-world development of animal embryos works normally. . . . The keys for developmental success are cellular mechanisms that provide robustness and buffer embryos from the environment and regulatory pathways that alter the developmental path in response to the conditions encountered. These mechanisms provide potent, although not impregnable, defenses against common stressors in development."[16]

If we imagine a recently fertilized fish embryo floating in the uppermost layers of the sea, its need for UV protection would certainly be greater than for its adult counterpart. Protection might come in the form of natural sunscreens by way of mycosporine-like amino acids, or through "behavioral" differences—eggs that drift to the bottom, out of harm's way—or by a ramped-up DNA repair system. Other embryos, faced with exposures to toxic substances, might produce high concentrations of efflux proteins, enabling them to pump out the offending chemicals.[17] However, note Hamdoun and Epel, "although embryos are well buffered for expected environment(s), rapid anthropogenic changes can overwhelm this intrinsic robustness. Of special concern are man-made chemicals that evade developmental defenses or misdirect developmental decisions."[18] Increased UVB levels, for example, well above those in life's recent history as a result of the atmospheric ozone loss, would likely pose a threat to buoyant embryos, despite their best defensive efforts. Additionally, says Hamdoun, increased defense sometimes comes with a price: "An important concept is that there can be trade-offs between defense and development, with selection sometimes favoring strategies that are advantageous for morphogenesis [growth and development] but render the embryo vulnerable to chemicals. Presumably that is one reason defenses are not expressed maximally throughout development and in all cells."[19] Embryo survival at the expense of the adult is not the most effective long-term strategy.

Sometimes conditions affect subsequent generations as well. This phenomenon, referred to as "adaptive tuning" by Hamdoun and Epel, is one of the more striking and insidious consequences of gene-environment interactions. The process results in environmental imprinting and, because it influences gene composition, can cause permanent and heritable changes.[20] The notion that the environment can influence genes in a heritable manner is rapidly gaining acceptance in the form of epigenetics, broadly defined as "the structural adaptation of chromosomal regions so as to register, signal, or perpetuate altered activity states."[21] Such induced adaptations in DNA can be revolutionary, leading to immediate and powerful changes. This capacity for overriding life's "blueprint" through epigenetic-induced novel phenotypes, observe Hamdoun and Epel, may be yet another evolutionarily preserved strategy. The potential for such rapid and lasting change may be one way for living things to deal with rare and dramatic environmental change—or not. Though the rapid phenotypic change may be advantageous under some conditions, Hamdoun and Epel write, "this generation of new phenotypes is a 'lottery approach,' meaning that most of these phenotypes are not adaptive. However, the gamble may provide a chance to escape from severe environmental bottlenecks."[22] Epigenetic responses occur universally across species, from bacteria to humans. Some of the more disturbing reports of epigenetic changes relate to the effects of endocrine-disrupting chemicals in second- and third-generation offspring of exposed females. Reduced fertility caused by maternal exposure to the antiandrogen vinclozolin, for example, has been found to persist across generations.[23] Only a decade or so ago, prior to recent advances in genetics, such an observation would have been rapidly dismissed. Epigenetics provides but one example of how much more we have to learn about the interactions between environment and genetic controls. Hormesis, as described in detail below, provides another.

Hormesis

Like epigenetics, hormesis is a highly conserved response to stress that is only now gaining the recognition it deserves. This concept received little attention throughout much of last century, and when it did, it was viewed more as a threat to toxicology than as a true response to chemical exposure. Then, in the latter part of the twentieth century, the molecular toxicologist Ed Calabrese turned his attention to investigating and then popularizing the phenomenon. After publishing volumes of papers, reviews, and books, Calabrese eventually convinced a reluctant audience that hormesis not only occurs but in fact may be a normal and common response to stress.[24] Hormesis, as defined by Calabrese, is "a modest overcompensation response following an initial disruption in homeostasis—that is, a type of rebound effect. The hormetic dose response therefore represents the effects of a reparative process that slightly or modestly overshoots the original homeostatic set point, resulting in the low-dose stimulatory response."[25]

The response provides a means of protection for cells residing in an environment where toxic chemicals (or other stressors) become increasingly available. Hormesis infers neither benefit nor harm. The result is complex, relative to the particular situation, and highly conserved: "Hormetic effects occur in essentially all plant, microbial, and animals species, affecting many hundreds of endpoints in numerous cell types and tissues, involving many hundreds of genes for each endpoint. The hormetic response represents a very basic and general strategy that occurs in all types of cells and tissue using a wide variety of integrative mechanisms."[26] Given what we know of life's ongoing exposure to environmental stressors, the evolution of a hormetic response is not surprising. Rather than allocating valuable resources and energy to protecting against inconsistent or persistent threats, life has essentially evolved an early warning system. The first inklings of exposure trip an early yet subtle response.

The epidemiologist Linda Gerber and colleagues offer another plausible explanation for hormesis: "Hormesis may be almost universal for substances normally present throughout geologic time, such as mercury. If mercury was always present in a tiny trace in the ocean and later habitats, organisms would have evolved ways of minimizing its damage at normal concentrations, and such adaptations may work moderately well at higher concentrations."[27] Though we know mercury can have devastating effects on the developing mammalian brain, very low concentrations of mercury were recently found to increase the number and growth rate of offspring in mallard ducks—a finding the authors attributed to hormesis.[28] Keep in mind that the end points of this study were number and growth. It is not known whether brain development or any other subtler end point was influenced. Nor is it known whether the cost of minimizing mercury's impact on the observed end points affected other regions of the defensive web. And of course, while some industrial chemicals induce hormesis, it is also likely that many do not.[29] If hormesis is a normal and first response, the lack of a hormetic response could render minute amounts of some industrial chemicals all the more insidious.

The defensive strategies discussed throughout this book evolved in response to environmental stressors or conditions present throughout much of life's history. But what happens when the chemical landscape changes dramatically, such as when chemicals are released, say, from a volcanic explosion? Or when a creature finds itself compelled to prey on unusual plants or animals in an act of desperation? Or when life is exposed to chemicals that are not only rare, but also novel? One hypothesis that is new to toxicology—attractor states—suggests that evolution has prepared life for the unexpected as well.

Attractor States: Preparation for a Rare Event?

Although well beyond the scope of this book, the adaptive immune response is considered the quintessential example of how living things prepare for the unknown. Yet there may be another, even more global response, which allows all living things to survive rare, potentially harmful events. Rather than any one specific response, I am referring to attractor states: a complex yet fascinating concept that I will do my best to explain, with some help from the molecular biology and biocomplexity scholar Sui Huang.

The responses within any one cell and the differences between individuals and species are complex, nonlinear, and dependent on both the external and internal environments. And yet it is becoming increasingly evident that "the space of environmental conditions is much larger than that of cellular response programs, there is not a program for each condition. . . . Thus, many environmental conditions map onto the same cellular response."[30] That is, each response to a specific toxicant (whether oxygen, metal, or UVB) requires some sort of signaling pathway, but there cannot possibly be a pathway for every stressor, particularly those rarely encountered throughout evolutionary history. As Sui Huang asks, "Why would molecules that cells have never seen interact in a specific way and elicit a stress response? How did such responses that preempt synthetic chemicals yet adequately protect cells from the latter evolve?"[31] It is a question that Huang and others suggest might be answered by considering the role of attractor states. This hypothesis is born of complexity theory, and the math underlying attractor states may be beyond the grasp of some of us—but the consequences are not.

Huang explains it as follows:

An attractor state is the product of the dynamics of a network of molecular interactions. [If] we speak of "quasi-energy,"[32] then an attractor state is a state of relatively lower quasi-energy, akin to the lowest point in a potential well (or valley in a landscape). It is the most stable state. The regulatory interactions between the genes, proteins, etc. assign to each cellular state (defined by the concentrations of all the proteins, metabolites, etc.) a particular "quasi-energy." The lower that energy, the more stable it is. For example, if gene A inhibits gene B, then all cellular states in which gene A AND B are both highly expressed would be of high quasi-energy, that is, very unstable, and the cell would move to the closest state in which gene A is high and gene B is low. Thus attractor states are locally preferred states in which regulatory conflicts in the network of interactions are minimized. This becomes very complex and abstract when we deal with tens of thousands of molecules instead of just two that regulate each other. Hence the metaphoric picture of a rugged quasi-energy landscape in which the valleys (energy wells) represent a large set of such preferred states—or attractor states—from which cells can "choose" is intuitively useful.[33]

This means that as cells alter their gene expression pattern, mathematically equivalent to movement on this landscape, they will be attracted to a particular stable phenotypic state (of the many available). This "best fit" is the lowest energy state that allows for a cell to survive in a given environment—whether it is a hormonal state of the tissue (during normal development) or a metal- or PAH-contaminated environment, or both.[34]

We have seen throughout this book that cells have evolved the capacity, through a complex network of signaling pathways, to switch from one state to another as they find the best "adaptive" state. That is, a cell that has ramped up its metallothionein or CYP response will eventually revert to a different state once the toxic threat recedes as a result of signaling controls. In the case of ongoing threats, cells may eventually achieve a state that represents the best fit for a particular situation. But what of the rare event? Of this incongruity between the universe of environmental conditions and the limited permutations of a cell's response programs, Akiko Kashiwagi and colleagues write, "It is unlikely that cells have evolved a specific signal transduction pathway for every environment it may encounter. Some attractor states may provide the optimal gene expression program for the cell to adapt to and cope with a particular, rare environment, yet no specific signaling pathway may exist that connects this rare external condition with the appropriate genetic program. . . . This raises the question: how do cells in the case of rarely occurring environmental changes switch to the adaptive attractor state of the network that expresses the appropriate genes?"[35] Based on evidence from yeast cells, Kashiwagi and others propose that rather than depending solely on highly evolved signaling pathways, cells may "randomly switch" attractor states and, should they hit on an adaptive state that best fits the situation, they may remain there even without the benefit of signaling.[36] In this way, yeast cells may have "a sort of Darwinian preadaptation for the evolution of signal-specific transduction pathways when a particular new environmental condition becomes dominant and hence contributes to evolvability."[37]

The attractor state hypothesis suggests that cells have the capacity for nonspecific switching between states—changes that are not dependent on time-tested chemical signaling pathways. This theory is still in the early stages of development. Yet Sui Huang argues that it may explain why cancer cells all too often elude one chemotherapy agent after another. Huang writes, "Attractor switches thus are a quick adaptation, a physiological process, that does not require random mutation of the genome and selection *a la* Darwin."[38] Should the hypothesis gain traction, it could spur further research into how attractor states relate to life's response to synthetic industrial chemicals. Whether "normal cells" have the potential to resist chemicals, particularly chemicals new to life, is currently anyone's guess.

Natural Defense

Throughout history, humans have tried to model nature's systems, forms, and functions—whether the flight of a bird, the tensile strength of a spider's web, or the gene delivery system of a virus. We are now just beginning to understand another ancient system: natural defense. Evolution has provided life with a robust, multipronged approach to chemical defense, from a raft of receptors sensitive to incoming chemicals to nonspecific and global stress responses that are highly networked. Yet humans will no doubt continue to alter the earth's chemical environment, challenging life's chemical defense networks. Whether these chemical additions will fly harmlessly under the radar, or cause subtle and insidious impacts on subsequent generations, remains to be seen.

PART 3

Human

Then

Chemicals toxic to life have been a powerful selective pressure since the origin of life

Toxic Evolution

Some of these chemicals may result in standing genetic variation or other mechanisms, which provide an enhanced potential for rapid evolutionary changes occurring within a few generations

Now

Species from bacteria to vertebrates to plants are all known to evolve in contemporary time in response to industrial chemicals

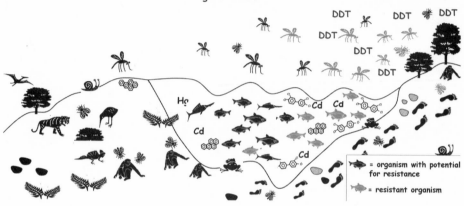

= organism with potential for resistance

= resistant organism

Contemporary evolution.

Chapter 9

Toxic Evolution

It is likely that most toxic chemicals in the environment will affect evolutionary processes. . . . We predict that a new field, evolutionary toxicology, will emerge to address these issues.

John Bickham and Michael Smolen

Whereas it is true that evolution has clearly not rescued all species or populations from extinction, explorations of extinction probabilities based on the limitations of selection and response stand in stark contrast to a growing literature demonstrating that surprising amounts of adaptive evolution occurs in the wild and laboratory within a human life span.

Michael Kinnison and Nelson Hairston Jr.

In short, humans cause particularly dramatic changes in organisms — and these changes are probably often adaptive.

Andrew Hendry and colleagues

When my daughter developed an infection that had begun streaking up her leg, my first thoughts were of methicillin-resistant *Staphylococcus aureus*, or MRSA. Although it was staph, fortunately it wasn't MRSA, and the antibiotics she had been prescribed won out. But with antibiotic resistance on the rise, we can no longer assume that common bacterial infections will be conquered with a simple course of antibiotics. For too long we believed we could easily outwit bacteria with antibiotics, and we did for nearly half a century before realizing our folly. We have ignorantly underestimated life's capacity for evolution. But there were early warnings. Back in 1945, when Alexander Fleming, father of modern antibiotics, won the Nobel Prize for discovering penicillin, he cautioned that inappropriate use of the antibiotic might have some unwelcome consequences:

> Penicillin is to all intents and purposes non-poisonous so there is no need to worry about giving an overdose and poisoning the patient. There may be a danger, though, in underdosage. It is not difficult to make microbes resistant to penicillin in the laboratory by exposing them to concentrations not sufficient to kill them, and the same thing has occasionally happened in the body. The time may come when penicillin can be bought by anyone in the shops. Then there is the danger that the ignorant man may easily underdose himself and by exposing his microbes to non-lethal quantities of the drug make them resistant. . . . Moral: If you use penicillin, use enough.[1]

In time, Fleming's scenario has come true for a variety of microbes and a number of antibiotics—including methicillin, the drug of choice after penicillin began to fail.[2] Although he could not have known the genetic mechanisms involved, he understood the implications: that life has a capacity for fast-paced evolution, and that persistent exposure to toxic chemicals has a role to play in this process. Today, evolutionary changes are taking place in species from bacteria to fish. These changes are occurring within our lifetime, and often as an incidental consequence of our activities. The notion that evolution can occur in contemporary time is quickly working its way into the evolutionary canon.

One of the most celebrated, and for a time controversial, examples of evolution in contemporary time is the peppered moths of industrial London. If there is an image burned into my memory from introductory genetics classes, it is of two moths clinging to a darkened tree trunk—representatives of the same species, one is dark and barely visible, while the other is light colored, flecked with black, and easy to spot. The industrial (or peppered) moths were the quintessential example of both natural selection and the impact of human activity on the natural world. Over the years, as soot from local industries began darkening London's tree trunks and branches, moth coloring adapted from light to dark. Following decades of environmental regulation and improvement, their coloring eventually reverted back.[3] For many of us, these moths were an introduction to the wonders of evolution, yet they were presented as an odd case of evolution induced by human activity. Unfortunately, they now have plenty of company, as an ever-increasing number of bacteria, insects, fish, and other vertebrates are now known to have undergone evolutionary change in contemporary time (decades, as opposed to hundreds, thousands, or millions of years) in response to our actions.[4]

Evolving Ideas

When I first learned about those peppered moths, evolutionary theory was based on the concept of gradualism. Yet over the years, understanding of the tempo and mode of trait change in nature has come to include alternative views about the pace of evolution. We were taught that the process of macroevolution, or speciation and adaptation, was extremely slow, taking place at imperceptible rates over thousands of years or more. Evolution plodded along, occasionally leaving behind traces in the fossil record. Transitions between forms were thought to represent prolonged gaps in the record that might one day be filled. That perception of gradualism received its greatest modern challenge in the early 1970s, when Stephen Jay Gould and Niles Eldredge proposed that instead of continual gradual change, species evolved through an "accumulation of discrete speciation events," combined with periods of stasis. And, they suggested, rather than simply a "passive response to unaltered environments," maintaining stasis was an *active* process.[5] In other words, stasis reflects a real evolutionary phenomenon and not just missing data from the fossil record. They referred to this as "punctuated equilibrium," which was at the outset controversial, misunderstood, and distorted by creationists—before eventually coming of age and adding to the prevailing evolutionary theory.[6] Today, though on a different scale, a similar reconsideration of the rate and manner in which microevolution, or small-scale changes occurring within a species, is taking place in the context of the study of "contemporary evolution."

Though the peppered moths provided only one example of human activity's influence on insect evolution, they were harbingers of things to come. Throughout the last century, as humans increasingly relied on pesticides and antibiotics to produce food and maintain human health, the capacity to rapidly evolve, at least in pests and pathogens, became not just clear, but also predictable. How far-reaching across taxa was this phenomenon? Until the end of the twentieth century, most observations of contemporary evolution (often referred to as "rapid" evolution) suggested that these were exceptions, and that they most often occurred in species with short generations and large broods. Yet there were few well-studied examples in vertebrate species. Unlike invertebrates, the long generation times and limited number of offspring produced by most vertebrates probably steered researchers away from serious consideration of evolution in many species. Studies conducted with bacteria may take days to weeks, while those with insects may extend over weeks to months. Demonstrating evolutionary change in many vertebrates takes years to decades. But this is not the case with all vertebrates, and a recent global history of species introductions and invasions has afforded quite a few seminatural experiments to help us understand how many vertebrate species evolve in the face of changing environments. Moreover, a number of long-term data sets of vertebrate populations came to fruition in the latter part of the last century.

One of the most powerful studies documenting evolutionary change over "contemporary time" in vertebrates is the thirty-plus-year study of Darwin's finches conducted by the evolutionary biologists Peter and Rosemary Grant.[7] In their now-classic 1995 and 2002 papers, the pair charted the course of natural selection and the evolution of beak shape and body size in a population of Galapagos finches. Their efforts revealed both predictable evolution on a generational scale and "evolutionary unpredictability on a scale of decades."[8] The power of such a long-term study toward illuminating the rapidity of contemporary evolution became clear, as the Grants themselves noted, in realizing that had they recorded beak and body size only at the beginning and end of the thirty years, they would have missed changes that occurred in the middle. "Natural selection" they wrote, "occurred frequently in our study, occasionally strongly in one species and oscillating in direction in the other."[9]

The evolutionary biologists Andrew Hendry and Michael Kinnison popularized and defined the term "contemporary evolution," in their popular 1999 article "The Pace of Modern Life: Measuring Rates of Contemporary Microevolution," as microevolutionary changes taking place in "recent times and on short time scales (less than a few centuries)."[10] Focused on rates of microevolutionary change and drawing on data from studies indicating contemporary evolution in vertebrate populations (including the Grants' study), the pair concluded that "evolution [as] hitherto considered 'rapid' may often be the norm and not the exception. We suspect that when populations or species are exposed to changing environments . . . evolution will appear rapid relative to that documented over longer time frames, or in undisturbed situations. Thus, claims of rapid microevolution should not necessarily be considered exceptional, and perhaps represent typical rates of microevolution in contemporary populations facing environmental change."[11] These contemporary populations include the antibiotic-resistant bacteria, pesticide-resistant insects, and metal-resistant worms discussed in chapter 4, and, as we will see, PCB- and dioxin-resistant fish.

Although rapid change can occur across species given similar types of selection pressures (a new antibiotic or pesticide), the genetic mechanisms underlying contemporary evolution can be quite different between a bacterium and a member of the animal kingdom. There are tremendous variations between species like bacteria and Darwin's finches when it comes to generation time (on the order of minutes for some bacteria) and population size (bacteria populations can blossom into the millions or billions in a very small space). Clearly, bacteria are in a class of their own — or at least a class very different from the rest of us multicellular eukaryotes — when it comes to maintaining the fidelity of genetic backgrounds.[12] As mentioned briefly in earlier chapters, bacteria have a greater capacity to swap DNA within and among populations through lateral gene transfer. That ability, along with a potential for altered DNA fidelity in the face of environmental stress and the existence of "mutator strains," has contributed to the notion that one day bacteria may indeed inherit the earth. Plus, some of the same mechanisms underlying the evolution of resistance (altered metabolic activity, for example) allow bacteria to use pesticides and other toxicants to their advantage. Consider one of the more recent environmental disasters: the BP oil spill. Bacteria metabolized extraordinary amounts of oil, turning our wasted energy into theirs and benefiting us both. This capacity to constantly adapt reflects, in part, the fluidity or infidelity of bacterial DNA. To some extent, the earth's bacterial genome is available to all strains of bacteria, a feature that can make it difficult to separate one strain or even one species from another.[13]

While such variability works well for single-celled life, more complex species engender greater investment in both individuals and populations. Multicellular life required genetic continuity and fidelity within individuals and species. Fortunately, there is more than one route to flexibility or adaptation in the face of environmental challenge.

Rapid evolutionary change, particularly in species with longer generation times and few offspring, may also occur by way of standing genetic variation or the presence of alternative forms of a particular gene (or multiple alleles) within a population.[14] Depending on a species' evolutionary history, older alleles, like ghosts from past environmental experiences, may be retained as gene variants that have managed to pass the test of time. As such, they may lead to faster evolution and to more rapid fixation of "tested genes."[15] Standing genetic variation likely accounts for the Grants' finches' morphology and many documented cases of resistance in other vertebrate species (and perhaps even in some insects). By virtue of their large populations and rapid generation time, insects *may* also, like bacteria, benefit from some amount of new or de novo mutation in addition to existing variation.[16] And at least one case—adaptive rapid change in deer mouse coloration—has also been attributed to a new mutation.[17]

While the Grants' long-term study might set the standards for identifying cases of contemporary evolution, not all researchers have had the opportunity for ongoing studies of such historical proportions. Nonetheless, observations of contemporary evolution have been recognized in an increasing number and diversity of species. And many of these observations have occurred in highly contaminated sites—indicating contemporary evolution in response to industrial chemicals or other human-induced, anthropogenic changes.

Plasticity or Natural Selection?

Back in the mid-nineties, reports of killifish (*Fundulus heteroclitus*, a minnow-like fish) resistant to PCB and dioxin toxicity piqued the interest of those of us studying the effects of these chemicals in wild fish populations along the eastern coast of the United States.[18] The killifish seemed to thrive in highly contaminated sites. Yet other species, including lake trout, were driven nearly to extinction in Lake Ontario, likely because of dioxin toxicity.[19] Similarly, local populations of mink, also highly sensitive to dioxin and PCB toxicity, may have met their demise by subsisting on a diet of PCB-contaminated fish.[20] Curious about the nature of killifish resistance, the aquatic toxicologists Adria Elskus, Anne McElroy, and I designed a study that we hoped would illuminate one basic issue: the inheritance of resistance. We knew that the CYP1A enzyme (discussed in chapter 6), a common marker of exposure to dioxins and similar chemicals, was relatively nonresponsive in some resistant killifish populations compared with the response in fish collected from less contaminated sites. Yet whether the behavior of CYP1A contributed to resistance or was simply a marker of resistance was unknown—as were the underlying genetics of resistance. Did resistant fish simply possess greater variation in CYP1A and associated genes, or was resistance a product of natural selection, or both? Had contamination resulted in evolutionary changes over the relatively brief period of time (some fifty or so years) in which their habitat had been contaminated by these industrial pollutants? Or had more ancient evolutionary processes provided them with the capacity to respond to what are now industrially relevant chemicals, which are either the same or similar to existing natural chemicals? (Given the universe of all possible organic chemicals, it may be that even as humans synthesize chemicals, there really is very little "new" under the sun.)[21]

Our study would test the responsiveness of both adults *and* offspring from a contaminated site located near Newark Bay, New Jersey (a population whose offspring were previously identified as resistant to dioxins), and a reference site, Flax Pond, New York, to environmentally relevant concentrations of PCBs.[22] In this way, we could at least determine if offspring shared characteristics with their parents. After collecting fish from both sites, and allowing six months for metabolism and excretion of the majority of PCBs (to reduce any effects of their home environment), we exposed adult fish to a mixture of PCBs. As expected, the adults were resistant to CYP1A induction; so too were their offspring.[23] Though the results suggested inheritance, a single generation was insufficient for ruling out direct influences of PCBs, other chemicals from the parents' native habitat, or some other unknown environmental influence on the parents that could be passed on to the offspring. Any declaration of evolutionary adaptation, however, would have to wait for further study. That was sixteen years ago.

While our study was suggestive, the observed resistance may also have been an expression of existing characteristics within the population. Sometimes an individual or its offspring can change phenotypic expression of form, coloring, or enzymes in response to environmental stress without changes in gene frequency. These changes can occur as a result of phenotypic plasticity, defined as "a single genotype to exhibit variable phenotypes in different environments. . . . Plasticity is physiological, but can manifest as changes in biochemistry, physiology, morphology, behavior, or life history. Phenotypic plasticity can be passive, anticipatory, instantaneous, delayed, continuous, discrete, permanent, reversible, beneficial, harmful, adaptive or non-adaptive, and generational."[24] Based on this definition, discerning the difference between adaptive changes attributed to plasticity versus contemporary evolution, even when observed across generations, can be tricky. Had we simply observed some manifestation of plasticity? And if we are just interested in life's capacity to respond to PCBs and similar chemicals, does the underlying mechanism really matter? Making the distinction, writes Michael Kinnison, depends on one's goals:

> If the goal is to actually talk about heritable differences between populations or heritable changes acting over time, then it is often important, since plasticity could fool you into thinking that a trait evolved when in truth the changes observed are largely due to direct environmental effects on trait expression. Conversely, if one is interested in just knowing how phenotypes influence performance (i.e., fitness) then it is not so important to distinguish plasticity and contemporary evolution. In reality, most cases of trait change in nature are likely to involve some elements of both plasticity and evolution. Indeed, isolating the two is a bit of an artificial construct based on the experimental approaches that scientists use to measure each. In reality, most traits are plastic to some degree (i.e., individuals can produce a range of phenotypes across some range of environmental exposure) and as such contemporary evolution probably often involves evolution of plastic responses to particular environmental conditions. You could think of plasticity as a trait in and of itself, one that is heritable and adaptive in allowing organisms to produce different phenotypes under environmental conditions that selectively favor different trait expression. That range of plastic responses for a given genotype is called a norm of reaction or reaction norm and it is evolvable and subject to selection to the extent that producing the right phenotype matters to survival and reproduction in a variable environment.[25]

Discerning the nature of PCB resistance would elucidate how some species endure ongoing exposure to at least one class of toxic chemicals, and why one population might survive and another go extinct. Because killifish reside along much of the highly contaminated eastern seaboard, and because they are so resilient, they have become the focus of many research groups. Over the two decades since our study, the aquatic toxicologist Diane Nacci and colleagues have revealed a great deal about the nature of killifish resistance to environmental contamination by PCBs, dioxins, and similar chemicals. Through a series of experiments involving fish collected from a number of contaminated sites (including Newark Bay), Nacci's group demonstrated not only that resistance is heritable across multiple generations, but also that it indicates evolutionary change rather than plasticity.[26] In one massive study, Nacci tested the responsiveness of first- and second-generation offspring of adults collected from twenty-four sites located along the eastern coast, from Virginia to Maine. After maintaining the different adult populations in the laboratory under the same conditions (referred to as "common garden" conditions),[27] for anywhere between six months and two years, Nacci collected and exposed embryos to a single dioxin-like PCB (PCB 126) while allowing others to mature over the following year so that *their* offspring could then be challenged with the PCB. This design reduces the potential for local environmental influences on successive generations. The common garden, writes Kinnison, allows the researcher to "largely nullify plastic effects on phenotypes. This gets around the problem that two genetically identical populations might appear different just because they are exposed to different environmental conditions."[28] Nacci's results provide a striking demonstration of heritable resistance, which, along with that of others over the past two decades, firmly established that killifish resistance reflects contemporary evolution induced by recent pressures of industrially produced chemicals. Additionally, because killifish from diverse populations all responded similarly in terms of resistance, these studies also provide a fascinating example of widespread convergent evolution: similar phenotypic changes occurring in distinct or unrelated populations.[29]

Establishing evolutionary change as Nacci and others have done with killifish is one large step in understanding the impacts of chemical contamination and life's capacity to respond. Determining the underlying genetic mechanisms places these observed phenotypic responses in the context of the greater genomic network (discussed in the preceding chapter). And while neither Nacci's group nor others have yet identified the molecular underpinnings of resistance, a recent analysis of gene activity in killifish by Nacci, the evolutionary biologist Andrew Whitehead, and others sheds some light on the broader implications of resistance. In addition to revealing the degree to which these evolutionary changes have truly converged (in terms of gene activity), their work indicates similar changes in both global and specific gene expression across resistant populations.[30] Using a similar experimental design as described above, but with offspring from just two different populations, the group focused on the transcriptome, the small set of RNA transcribed from DNA. This provided insight into which genes are turned on or off—the up-regulation and down-regulation of genes. The differences between resistant and nonresistant embryos following exposure to PCBs are striking. While the RNA produced by sensitive embryos showed massive up-regulation and down-regulation involving nearly two thousand genes (including many involved directly and peripherally with the CYP response), the resistant population's response involved only around five hundred gene transcripts, most of which were up-regulated. Observing the activation of AhR target genes in the sensitive population, but not in the tolerant population, Whitehead and colleagues suggest that the mechanism of resistance most likely involves blockage of the AhR pathway.[31] Based on the large differences in transcriptome responses, this finding could have far-reaching consequences. The large differences in the numbers of genes involved in responding to PCBs is unsurprising considering the highly networked nature of the environmental response genome in general (discussed in the previous chapter) and AhR in particular (as discussed in chapters 7 and 8).[32] Although we tend to focus on easily observable phenotypes (like responsive or nonresponsive), there are likely a number of other characteristics of resistant populations that we have yet to identify.

Killifish aren't the only PCB- and dioxin-resistant fish species. The molecular toxicologist Isaac Wirgin has also discovered that Atlantic tomcod, a species of bottom-dwelling fish inhabiting the PCB-contaminated Hudson River, have also developed resistance.[33] Teaming up with Mark Hahn, Wirgin and others have solved at least part of the mechanism puzzle in tomcod. Resistant tomcod, it turns out, are lacking two amino acids in one of the two AhRs in fish (AhR2). This mutation appears to *indirectly* affect the receptor's ability to bind with contaminants like PCBs, PAHs, and dioxins.[34] Summarizing their work, Wirgin and colleagues conclude, ". . . Hudson River tomcod have experienced rapid evolutionary change in the 50 to 100 years since the release of these contaminants [halogenated aryl hydrocarbons like PCBs and dioxins]. Our results indicate that resistance can be due to one structural change in the coding region of a single gene, and that evolutionary change in anthropogenically challenged natural populations can be rapid."[35]

Back in 1994, just as our killifish studies were getting underway, the geneticists John Bickham and Michael Smolen, recognizing the potential for toxicants to impose selective pressures on populations, suggested a new discipline: evolutionary toxicology. This subdiscipline of toxicology, wrote Bickham and Smolen, "will deal largely with the emergent effects of environmental toxins. That is, evolutionary effects are changes at the DNA sequence level that are not necessarily the direct result of a mutation inducted by the pollutant. Rather, they are the result of organisms adapting to a polluted environment and thus are fundamentally different in nature, and emergent from, lower-level processes such as ecologic effects and toxic effects."[36] Almost two decades later, Bickham set forth the four cornerstones of evolutionary toxicology: (1) genome-wide changes in genetic diversity; (2) changes in allelic or genotype frequencies caused by contaminant-induced selection acting at survivorship loci; (3) changes in dispersal patterns or gene flow, which alter the genetic relationships among populations; and (4) changes in allelic or genotype frequencies that are caused by increased mutation rates.[37] Implicit in this approach is the understanding that a chemical that binds to the estrogen receptor or to the AhR may have farther-reaching consequences than those observed in a single- or even multiple-generation experiment.

The capacity to evolve rapidly could be considered one of the most dramatic defensive responses to toxic chemicals. While more likely to occur in some species than others, it is clearly more common than previously imagined. Though contemporary evolution may be advantageous in the short run, it is unclear how well rapidly changing species will fare in the long term. A single altered gene could have broad influences across the genome or on the function of key enzymes. (While AhR affinity for dioxin may be reduced, what of its natural or endogenous ligand?) Plus, what happens to these species when environmental conditions change again—perhaps as contaminated sites are cleaned up? Hendry and Kinnison suggest that rapid evolution in some cases may become unsustainable, leading eventually to extinction—nature's version, perhaps, of *damned if you do, damned if you don't*.[38] The capacity to evolve over contemporary time periods must be balanced by rates rapid enough to enable survival, yet not so rapid that they exceed sustainability. This brings us to the last section of this chapter, where we consider the cost of rapid evolution. Just as a cell or body must maintain some internal homeostasis, living things must maintain a balanced relationship with their surrounding environment. When a cell metabolizes potentially harmful chemicals, there is often a cost: the production of reactive oxygen as a by-product, or the use of other valuable resources to produce the enzymes necessary for metabolism. When considering rapid evolutionary changes, we might similarly ask if there is a cost in terms of fitness—the relative capacity of one organism to pass on its genes to the next generation.

Nothing in Life Is Free

The array of genes turned on and off in Whitehead and Nacci's killifish populations begs the question of cost: What is the consequence of such massive changes in gene expression? Is there a trade-off in terms of fitness or survival? It is a question Nacci and others have thought about ever since identifying resistant populations. Of the potential cost to her killifish, Nacci writes:

> The tolerance mechanism that seems consistent across tolerance in killifish is a poorly responsive AHR pathway. Bear in mind that this a "null" mechanism, whereby function is poor or lost, in other words, no predicted costs from protein upregulation, like those associated with some insecticide resistance mechanisms. Specifically, this "down-regulation" is protective against the acute toxicity of dioxin-like compounds but it may enhance the toxicity of compounds whose detoxification is dependent upon the AHR pathway. . . . Empirically, few costs have been identified for the tolerant killifish populations (at least among northern tolerant populations), except for the "mechanistic" or loss of detoxification cost described above. Other hypothesized mechanistic costs are based on perceived (but poorly known) relationships between AhR and other systems (immune, endocrine) or AhR's role in absence of xenobiotic agonists, for example, in normal development. But these have generally not been observed. Another generally

predicted cost is related to expected loss in genetic diversity following intense selection—as would seem to have occurred to produce tolerant populations. But none of the populations have low diversity. . . . In terms of costs, it may be interesting to consider that the contaminant-binding AHR pathway has evolved in vertebrates, yet we don't think of the "cost" of its absence in invertebrates, which generally enjoy the benefit of being less sensitive to PCBs and similar contaminants than vertebrates.[39]

Nacci and others have yet to document the cost of resistance to killifish populations, although Wirgin and colleagues concede that some cost to resistant tomcod is likely; along those same lines, no cost has yet been identified with resistance to metals in the marine worms discussed in chapter 4.[40] However, it is widely known that some forms of pesticide resistance are costly to the beneficiary—not to mention to society. The consequences of pesticide resistance are most clearly revealed by organophosphate (OP) resistance in mosquitoes. Forty years of pesticide application has yielded not only a worldwide plague of resistant mosquitoes but also an unprecedented opportunity to observe the mechanisms of evolution. Organophosphate pesticides inhibit the normal metabolism and removal of the neurotransmitter acetylcholine by interfering with acetylcholinesterase (AChE), the enzyme responsible for its metabolism.[41] If not removed from its site of action, the buildup of acetylcholine (the normal substrate for AChE) causes a range of effects, including uncontrolled smooth muscle contractions, twitching and paralysis of skeletal muscles, and changes in sensory nerve function and respiratory depression. OPs compete with acetylcholine for binding with AChE. In some cases, an OP pesticide irreversibly decommissions the AChE. In either case, OP exposure leaves victims to the mercy of their own neurochemicals. One form of resistance to OPs involves the *ace-1* locus, which codes for AChE. Resistance to OPs are conferred through an allele, *ace-1^R*, which contains a single amino acid mutation in the AChE. An enzyme with *ace-1^R* has a lower affinity for OPs—a life-saving benefit in the presence of OP. But the altered enzyme also has a lower affinity for acetylcholine, resulting in an enzyme that is up to 60% less efficient. In the absence of OP pesticides, resistant populations are faced with an acetylcholine buildup, just as if they had been exposed to OPs.[42]

It is well documented that environmental stresses can cause mutations that confer resistance, resulting in cost-benefit trade-offs like those described above. According to the evolutionary biologist

Pierrick Labbé and colleagues, whose research focuses on pesticide resistance and the evolution of the *ace-1* gene in particular, the amelioration of negative effects by more beneficial mutations creates a sort of "evolutionary inertia." Yet once selection pressures are removed, evolution can reverse course.[43] And the evolution of resistance is far more complicated than simply maintaining balance between "good" and "bad" mutations. Seldom does a mutation insert itself with laser-like accuracy. Given the networked nature of life's genomes, a single alteration will likely have multiple consequences. Sometimes the selection of resistant alleles carries with it selection of other alleles in the form of selective sweeps. Alleles that are physically linked may be influenced by direction selection, even if all are not under directional selection pressure. This is, in essence, a reduction in variation through association. In the case of insecticide resistance, such sweeps may provide resistance to one toxicant while diminishing the capacity to detoxify others.[44] Other times, the costs are directly associated with the benefit—as in the case of the resistance conferred by *ace-1* in mosquitoes. And sometimes, a seemingly simple mutation like *ace-1* can, write Labbé and colleagues, "trigger the evolution of the genetic architecture and gene number."[45]

Because *ace-1*-mediated resistance comes with such a high price tag yet is critical for surviving OP exposure, evolution has provided resistant mosquitoes with a "work around," in the form of gene duplications. Having multiple genes, as discussed in previous chapters, allows an option for change without losing function. In this case, gene duplications have provided mosquitoes with four "alleles" for both the resistant and the sensitive alleles of *ace-1* (which Labbé refers to as *ace-1*^{D}). The heterozygous (or mixed) combination of alleles provides mosquitoes with an opportunity for resistance *while* maintaining appropriate levels of acetylcholine.[46] Labbé and colleagues conclude, "In the long term, selection may produce exquisite adaptations, but this study lays bare that even impressive adaptations are likely to have begun with a process of trial and error that seems to be anything but optimal. It appears that natural selection is forced to tinker with available variability, despite the costs, rather than build impressive and cost-free adaptations that are wholly novel."[47]

Oftentimes, as we have seen throughout this book, life's response to toxic chemicals is robust. Yet just as "history is written by the winners," we tend to focus only on the survivors. Why killifish and tomcod adapted, survived, and possibly even thrived in PCB- and dioxin-contaminated environments while other species like trout and mink succumbed is a question that must be answered if we are to fully understand life's response to toxic chemicals. Aside from interspecies differences in the acquisition of resistance, many other issues remain. Would resistance evolve in animals exposed to high concentrations of multiple contaminates? What kinds of concentrations are required before resistance is observed? How might a resistant species affect other organisms with which it interacts?[48] Consider that resistant fish may accumulate large concentrations of toxicant, which may then be passed on to predators. And how might evolutionary history affect a species' ability to evolve? Although we know that bacteria, insects, and some fish species rapidly evolved the capacity to resist a multitude of industrial chemicals, other species, like the Grants' finches, evolved in response to naturally occurring environmental factors: changes in food supply, drought, or climate swings. For the vast majority of multicellular species, standing genetic variability, the product of a population's evolutionary history, will be critically important in determining a species' capacity for contemporary evolution.[49] In the past, life did not have to contend with DDT, PCBs, or high concentrations of mercury, lead, or chromium. Many synthetic chemicals are nothing new when it comes to basic chemical structures, yet as chemistry advances, so too may our capacity to alter chemicals in novel ways. Developments like nanochemistry (the synthesis of chemicals on a nano-scale, as discussed in the next chapter) will likely present entirely new challenges. Understanding how life might respond to the relatively abrupt changes we introduce, and the long-term consequences of those responses, is the challenge for toxicologists, environmental managers and regulators, geneticists, evolutionary biologists, and ecologists alike. As Hendry and Kinnison observed, "Perhaps the greatest contribution that evolutionary rate estimates will ultimately make is an awareness of our own role in the present microevolution of life and a cautious consideration of whether populations and species can adapt rapidly enough to forestall the macroevolutionary endpoint of extinction."[50] In the next and final chapter, we consider the effects of industrial chemicals on populations that have yet to show signs of adaptation.

Then
Over the eons, as Earth and life coevolved, life was exposed to an ever-changing palate of physical and chemical toxicants

Toxic Evolution
In turn, through evolution life evolved the capacity to neutralize and detoxify a diversity of natural toxicants, most of which we continue to benefit from today

Now
Yet we are changing Earth's environments and introducing unprecedented challenges to these protective systems

Modern chemicals now challenge ancient defenses.

Chapter 10

Toxic Overload?

Complex morphological or life history traits that depend on many generations are less likely to evolve quickly in small-population, long-generation species. These are the species, including most of the large animals and plants, that are most at risk of extinction due to poor ability to adapt evolutionarily to global change.

Stephen Palumbi

The final chapter of this book is truly a work in progress. How *will* life's toxic defense mechanisms respond to industrial age chemicals? We are the products of an ancestral line that survived the sun's intense ultraviolet light, the earth's metals, the poisons of plants and animals, and even life's own renegade cells. Whether one takes the long view of evolution or a contemporary view, naturally occurring chemicals have profoundly influenced evolution. As a result, our lives depend on redundant detoxification systems, membranes studded with protein pumps, filter-like organs, and layers of protective skin. Yet in today's industrialized world, these defenses are challenged in ways unlike any in the past. The opportunity for an evolutionary mismatch is not a matter of *if*, but *when*.

Life has existed on Earth for over three billion years; modern humans have been here for less than three hundred thousand years. Yet within our incredibly brief tenure on this planet, we have transformed it to the extent that our impact may join other major events in life's history, like oxygenation, asteroids, and ice ages, as a cause of major extinctions. Extinction is common; in fact, without ongoing extinction and speciation, we wouldn't be here. According to Anthony Barnosky and colleagues, major extinctions of the past share several features in common, including rapid rates of loss and the eradication of over 75% of existing species: "Hypotheses to explain the general phenomenon of mass extinctions have emphasized synergies between unusual events. Common features of the Big Five suggest that key synergies may involve unusual climate dynamics, atmospheric composition and abnormally high-intensity ecological stressors that negatively affect many different lineages. This does not imply that random accidents like a Cretaceous asteroid impact would not cause devastating extinction on their own, only that extinction magnitude would be lower if synergistic stressors had not already 'primed the pump.'"[1] If all else were in balance, perhaps our use of industrial chemicals would create nothing more than a blip, affecting only a few localized populations. But combined with climate changes, habitat degradation, spread of invasive species, and overuse of resources, our alteration of the world's chemistry could contribute to (and perhaps cause) the sixth major extinction. Conclude Barnosky and colleagues, "The huge difference between where we are now, and where we could easily be within a few generations, reveals the urgency of relieving the pressures that are pushing today's species towards extinction."[2]

A dire picture, yet we also know that life is resilient. Resiliency is in our genes, and much of it results from the challenge of maintaining balance amid unrelenting environmental stressors, from physical to chemical. Perhaps the metazoan strategy of multiple lines of defense—in addition to compartmentalization of defenses (protective skin, the filtering role of the liver, and in certain mammals, a multilayered placenta surrounding the fetus) and the capacity for evolution in contemporary time—*is* adequate. But what does it mean to be adequate? How much of a "hit" can a single cell, an animal, or a population take before it becomes irreparably altered? The DNA repair in our skin cells stems the damage caused by UV rays leaking through the atmospheric ozone layer. This system is and has always been fallible—but is there less room for error with increased need for repair? Then there is ground-level ozone. How vulnerable are our lung cells to ozone emanating from our cities and nearby regions? And as ozone attacks our lung cells, will their response to increased loads of microscopic particulates—which contribute to the production of reactive oxygen species—compound the damage?

At what point does life become overwhelmed? Global atmospheric changes have certainly been associated with major extinctions, and oxygen provides one of the oldest documented examples. Ironically, after most living things had not only developed protections against oxygen toxicity but also become dependent on the gaseous molecule, *reductions* in oxygen may have sparked one of the first widespread extinctions of complex animal life-forms around five hundred million years ago.[3] Our activities may pale in comparison to the rise and fall of oxygen, but we are, for now, the single most important purveyor of change to the earth's environments. We have made these changes in a blink in time. Though sudden climate change or habitat degradation may be more obviously detrimental, chemical contaminants could set the most vulnerable species on the path to extinction—and they could rapidly and directly affect evolutionary change in others.

Over a decade ago, a prominent group of ecologists and geochemists published an article in the journal *Science* titled "Human Domination of Earth's Ecosystems." In their opening paragraph, they wrote, "All organisms modify their environment. . . . Many ecosystems are dominated directly by humanity, and no ecosystem on Earth's surface is free of pervasive human influence."[4] Since its publication, this article has been cited more than three thousand times. Though focused on dramatic changes—including land transformation, degradation of marine ecosystems, and changes in biogeochemical cycles—the authors also highlighted the importance of synthetic organic chemicals. They noted that in the late 1990s, we produced more than "100 million tons of organic chemicals representing some 70,000 different compounds."[5] Since then, the manufacturing of all chemicals has increased by roughly 15% in the United States alone. In the 1950s, industrial plastic resin production was a mere 3.3 billion pounds; today, the world's plastics production is over 500 billion pounds, with 300 billion additional pounds added over the past two decades.[6] And the persistence of some of these chemicals (specifically, halogenated organics like PCBs, and DDTs and metals like mercury; for details, see the selected chemical profiles in the appendix) means that despite efforts to curtail their release, they continue cycling in the environment. In 2009, the Chemical Abstracts Service, which catalogs and tracks all known chemicals, announced the registration of its fifty-millionth "novel" chemical—the last ten million chemicals having been registered over the preceding nine months.[7] There are plenty more novel chemicals to be found. According to some estimates, the universe of stable organic compounds ranges anywhere from 10^{18} to 10^{200}.[8] Of the industrial and consumer-use chemicals, the toxicologist Thomas Hartung estimates that two thousand to three thousand have been studied extensively; three thousand to five thousand have been tested using rapid but limited testing protocols; eleven thousand are prioritized for testing; and upward of a hundred thousand are in use, in millions of synthesized substances and in an unlimited number of combinations.[9] It is a daunting situation.

Many of these chemicals have already made their way into living things, from deep-sea fish to polar bears to humans.[10] And although there is growing evidence of contemporary evolution in response to some chemicals in some species, the broader potential for these new chemicals to contribute to environmental selection pressures (whether acting alone or in combination with others, whether naturally occurring or human induced) is unclear. In addition to age-old metals like mercury and cadmium, chemicals that are now a part of life include the "legacy chemicals," like PCBs and dioxins, and current-use chemicals, like the perfluorinated compounds we depend on for our waterproof and nonstick products, the polybrominated chemicals that act as flame-retardants in plastics and foams, and the bisphenol A in plastics and other products. All of these are new to the earth's environments and life's chemistry. Predicting and quantifying the risks from long-term exposures to humans and other species, particularly intergenerational impacts, is one of the greatest challenges facing toxicologists, reproductive biologists, ecologists, managers, and regulators. Prevention is at the heart of chemical regulation. Yet it is a process that has been experiencing its own chemical overload for decades, as regulatory agencies struggle to incorporate cutting-edge analytical techniques while facing a backlog of untested or inadequately tested chemicals.[11] In an unprecedented letter published in *Science*, members of eight different scientific and clinical societies—from the American Society of Human Genetics to the Society for the Study of Reproduction—representing more than forty thousand members, expressed their growing concern over the inadequacy of federal chemical regulators to safeguard Americans from the impacts of chemical exposure: "Most, if not virtually all Americans, are exposed to contaminants in the environment that cause serious health effects in animal models. Direct links to humans remain uncertain, but there is sufficient experimental evidence to raise concern. Furthermore, there is growing evidence that some chemicals once thought to be safe and allowed into common and, in some cases, abundant commercial use may not be as benign as previously assumed."[12]

Some commercial chemicals will come and go, leaving little if any trace—even as they cause toxicity to individual members of a species. Others will leave their mark buried deep within the earth's soils and sediments. And some will leave their mark on life in the form of altered allele frequencies, the result of selective processes. Chemical testing and regulation have no doubt improved when it comes to protecting humans and wildlife from acutely toxic chemicals. Yet the more problematic chemicals are those that slip through unnoticed, causing subtle impacts on biological systems. We overlook the dangers of many chemicals because we fail to understand the biological relevance of the system with which they interact; or we are unable to predict how very small amounts might behave in the presence of other chemicals; or we focus on one response and neglect the networked nature of life's response to chemicals and other stressors. While predicting toxicity is an ongoing challenge, our ability to *detect* chemicals has greatly improved—to the extent that an ever-growing list of industrial use and consumer use chemicals are routinely measured in both human and wildlife populations.[13] Some of these chemicals have been banned for years, while others remain a large part of our chemical culture. In the appendix, I've included very brief profiles of a few select chemicals: PCBs, mercury, CFCs, endocrine-disrupting chemicals in general, and nanoparticles. Most have been discussed in earlier chapters and all are important—not only because of their presence in the environment and therefore in life itself, but also because they serve as an example of the risks we take by allowing the large-scale release of chemicals. These chemicals are, for now and the foreseeable future, a part of life on Earth.

Of course, these few examples are just the tip of the chemical iceberg. Consider atrazine, one of the most commonly used pesticides in the world (although banned by the European Union). There are a growing number of papers citing atrazine's subtle and not so subtle reproductive and developmental effects on frogs, and now fish, at environmentally relevant concentrations. A recent panel convened by the EPA concluded that "atrazine may affect hormonal milieu [in women], and possibly reproductive health outcomes."[14] Yet it remains the herbicide of choice in the United States and around the globe.

Then there are the flame-retardants known as polybrominated diphenyl ethers (PBDE). These chemicals are structurally related to PCBs, yet the use of PBDEs rose *after* the banning of PCBs. Like PCBs, these chemicals are now found throughout the world, in living things as far removed from industrial processes and consumer products as Arctic waterfowl, seals, and polar bears. A recent study suggests that PBDEs interfere with women's ability to become pregnant (i.e., their "fecundability").[15] To date, only certain members of the PBDE chemical family have been removed from the market.

Perfluorinated chemicals, used in nonstick and waterproof coatings, have also recently caused concern. These chemicals are long-lived, cycling around the environment in unexpected ways. High concentrations cause reproductive and developmental impacts in animal studies, yet it is unknown how the small amounts measured in human blood might affect health and development. Like PBDEs, their pervasiveness in the environment and in humans is enough to cause concern.[16]

Placing all these chemicals into an evolutionary context, there are many questions we might ask: What are the long-term effects? How will small amounts of these chemicals interact when they mingle in our blood, or within a liver cell or an embryo? How might combinations of these chemicals affect a fish or frog embryo as it also copes with temperature changes? Or alterations in food sources? Or a degraded habitat? Will underlying genetic changes provide these populations with greater resiliency to different threats in the future? Or will they cause a reduction in genetic variability, leaving them more vulnerable? And how might they affect longer-lived species with few offspring—like humans?

Advances in toxicology, chemistry, and related fields, combined with aggressive regulation, have led to a dramatic reduction in the indiscriminate release of industrial chemicals, at least in the United States and Europe. Now the challenge has changed from curtailing the obvious to ferreting out the insidious. Toxicology is no longer a "kill 'em and count 'em" science focused solely on gross effects—it hasn't been for decades, and lethality testing is on the decline. It is now a science charged with identifying subtle, multigenerational, long-term impacts of chemicals. The science of toxicology covers a broad spectrum. On one end, toxicologists seek to understand how chemicals interact at the cellular and molecular level. At the other end are those who strive to understand and predict the effects of chronic exposures to complex chemical mixtures on individuals *and* populations, from microbes to beluga whales. The two ends meet where receptors transport chemicals that turn on DNA, where enzymes are produced in response, and where networks in cell membranes respond in a coordinated manner: the mechanisms of the toxic response. It is here that incorporating evolutionary concepts may be most useful.

(R)evolution at Both Ends of the Spectrum

Throughout this book I have referred to the role of omics as a collection of methodologies providing toxicologists, molecular biologists, geneticists, and evolutionary biologists with the tools to dig deeper into the phylogenetic and evolutionary history of genes, proteins, and enzymes. How might toxicologists harness the power of these tools to sift through the overwhelming number of chemicals and identify those presenting the greatest threat, at least to humans? One method has gained a great deal of attention and momentum since it was first proposed in 2007 by the National Research Council's Committee on Toxicity Testing and Assessment of Environmental Agents and reported in *Toxicity Testing in the 21st Century*. The committee recommended a transition from whole animal testing, combined with a patchwork of in vitro tests for specific enzyme or receptor function, to an approach that reveals the network of responses by identifying and analyzing the mechanisms and pathways of toxicity (PoT).[17] It would be as if one could peer into a human cell and visualize the workings of its biological machinery in response to a chemical intruder. It is a vision requiring a shift in paradigms from defined end points to "identification of critical perturbations of toxicity pathways that may lead to adverse health outcomes."[18] According to committee member Melvin Anderson and committee chair Daniel Krewski, the report has sparked "a healthy and necessary discussion within the scientific community about the opportunity and challenges provided by the NRC vision for the future of toxicity testing."[19] The reactions, they observe, range from "extremely cautious, even pessimistic" to "guardedly optimistic." While some voiced concerns that the new technique must truly fit the needs of the field and that toxicologists and regulators must not place "overly high expectations" on new methodology, others emphasized the importance of appropriately defining adverse effects and the difficulties of translating in vitro effects to whole animals.[20]

Whatever the outcome of this particular effort to modernize the field, toxicology is at a crossroads. "The path forward," writes Anderson and Krewski, "will not be easy. It will require hard work, commitment to improving our current test methods, and an ability to make midcourse changes as scientific advances in toxicity testing are realized and the interpretive tools needed to evaluate new toxicity test data mature. The larger question is whether the effort is worthwhile. Our opinion on this remains unchanged. Toxicity test methods need to make better use of human biology and mode of action information to adequately assess risks posed to humans at relevant exposure levels."[21] One approach to improving mode of action will involve genomics combined with proteomics, or toxicogenomics. It is a methodology with the potential to combine the advantages of both reductionist and holistic approaches to chemical toxicity—like a living impressionist painting. Yet making the approach useful, write Daniel Krewski and colleagues in a summary of their NRC report, requires not only selecting key pathways but also a refined sense of the "patterns and magnitudes of perturbations" indicative of adverse effects.[22] Taking the approach one step further, Thomas Hartung and Mary McBride suggest aiming for a comprehensive catalog of these paths, including those key responses that are most sensitive and those that might provide some capacity to buffer and prevent toxic effects—in essence, the complete human toxome.[23] These pathways likely include some of those discussed throughout this book, which when in response mode or overwhelmed may be a stepping-stone along the path to toxicity. Throughout its existence, life has fended off toxic chemicals—so how many pathways have evolved over billions of years? "At the moment," observe Hartung and McBride, "any number is pure speculation. . . . Evolution cannot have left too many Achilles heels given the number of chemicals surrounding us and the astonishingly large number of healthy years we enjoy on average. How many PoT there are depends very much on the definition of PoT—what is a PoT on its own, what are variants, what are groups, etc.? Most likely we still focus too much on linear pathways. As we increasingly learn that processes in living organisms are networked we will likely learn that PoT are mostly perturbations of the network, not a one way chain of events."[24]

Acknowledging that a toxic response is the outcome of an overloaded network of "normal" responses, seeking out signals of these perturbations at the molecular or mechanistic level would, at the very least, provide insights into the complexities of life's response to chemicals. It would also present an opportunity to learn about the interactions between critical chemical combinations and cells of different type, age, and sex, and from different human populations. One way to understand the role of PoT, and perhaps to identify those which are key identifiers of chemical perturbations, is to seek out the evolutionary history of these pathways and networks, from original functions to the building and diverging of gene families to their current distribution across species. It is a tall order, yet one that is within closer grasp than ever before.

The discussion above belies the continuing tradition of toxicology as an applied science that has served regulators and managers for over a century. Since its origins, toxicology has (like life itself) diverged, duplicated, and evolved. Some of the earliest and most traditional branches included those devoted to the protection of human health, and so, by definition, the individual. This is the goal for the twenty-first century: to better protect individuals. It is a relatively straightforward charge in contrast to that of ecotoxicology. This field emerged in the tumult of the late 1960s and 1970s, following the publication of Rachel Carson's groundbreaking *Silent Spring*[25] and the advent of the Environmental Protection Agency. By definition, it must confront the complexity of life because it focuses both on individuals and on populations, communities, and ecosystems. One of the more important insights that population studies have afforded us is, as discussed in chapter 9, the influence of toxic chemicals on the evolution of exposed populations. As observations of contemporary evolution become more frequent, we must consider the most insidious consequences of our activities, including the large-scale release of chemical contaminants. The chemicals we have unleashed on the earth are causing genetic change, possibly on a large scale—and we are only just now grasping this reality. As the geneticist John Bickham writes, "Of fundamental importance, it must be emphasized that these genetic impacts are emergent effects not necessarily predictable by study of contaminant exposures or even the understanding of the mechanisms of toxicity, even though contaminant exposure is the root cause of the effects."[26] It is a humbling thought, particularly as toxicologists spar over the definition of adverse effect and shape the way for a new, bottom-up, mechanistic approach to toxicology. To borrow Richard Feynman's famous phrase, just as "there is plenty of room at the bottom,"[27] there is also plenty of room at the top.

We might do well to keep both ends in mind if we are to protect future generations of humans and wildlife.

Before concluding, I must acknowledge that there are many aspects of toxic defense and its evolution that I have either purposely or inadvertently neglected. As stated in the preface, my aim was to present a different context for thinking about toxicology, rather than to provide a complete literature review of all related ideas. Yet there is one very important emerging discipline that demands mention: green chemistry. If successful, this approach would reduce our chemical impact and perhaps make much of the above discussion moot. And it might benefit, if only in subtle ways, by looking to the evolution of life's responses to toxic chemicals. As a relatively new field popularized over the past two decades, green chemistry aims to design safer chemicals. (Those interested in learning more might begin with the review by the green chemistry pioneers Paul Anastas and Nicolas Eghbali, "Green Chemistry: Principles and Practice.")[28] Recent advances in toxicology, particularly the focus on of how living things interact with chemicals at the molecular and genetic level, in addition to the depth provided by studying the evolutionary history of these responses, will likely benefit practitioners of green chemistry.

A Conclusion Evolves?

As we have seen throughout this book, much of the research into the evolutionary roots of life's responses to toxic chemicals has emerged only over the past decade or two. This melding of toxicology and evolution reflects, in part, the revolution in genomics, the increasing curiosity of biochemical or molecular toxicologists about the origins of interspecies similarities and differences in the biochemical systems and defensive systems to which they've devoted their careers, and the growing awareness of the role of toxicants in contemporary evolution. Given the history of life's relationship with chemicals, one wonders how defenses that evolved for three billion years on a prehuman and preindustrial Earth will fare in this modern world. The easy answer is not well—particularly when we consider the impacts of large amounts of toxic chemicals on individuals, whether mollusks, fish, birds, or humans. The more difficult answer must address what happens when individuals are chronically exposed to small concentrations of the complex mix of chemicals we have released into this world. When DNA photolyase kicks into action in amphibians living at high altitude, what happens if these animals are exposed to the chlorinated pesticides and mercury that rain down from the atmosphere? If our CYP enzymes are increasingly metabolizing a variety of pharmaceuticals, what happens when we add one more, or change our diet, or breathe in chemicals like polyaromatics bound to micron-sized air pollution particulates? Or how might a fish exposed to estrogenic chemicals respond to subtle temperature changes? More important in the end, how might populations of marine worms, fish, birds, or humans respond? For nearly one hundred years toxicology has addressed these questions by taking a relatively top-down and piecemeal approach, moving from observations on the whole animal, to organs, to biochemical changes, and now, more recently, to genes. As toxicologists begin looking from the bottom up, their understanding of how these systems responded to chemicals and developed over time may allow them to integrate traditional techniques with newer approaches—and make them more effective.

It has often been written that we live in a sea of toxic chemicals. While this is true today, it was also true eons ago—with one glaring difference. Life on Earth is now subject to a virtual onslaught of chemicals associated in one way or another with human activity. To understand how life might respond to this unprecedented chemical milieu, we must explore how it evolved in the past. This book is meant only as a beginning. My hope is that all branches and levels of toxicology, from the classroom to predictive models, will eventually incorporate broad thinking about evolution. We are a society built on chemicals, and there is no turning back. Yet we can certainly improve how we produce, use, and release chemicals by striving for a better understanding of how they affect wildlife and human health. We have to do so. There is no higher ground, no corner of the earth where life can escape the influence of toxic chemicals. The choice must not be to "evolve or die."

The purpose of this section is to highlight the recent history of a few
key chemicals. It is not intended as, nor do I attempt to provide, a review focused on toxicity and current literature. There is no doubt that
all these chemicals are toxic at concentrations available to humans and
wildlife. The only exception is the broad category of nanoparticles
produced for industrial and consumer use, for which there currently
are not enough data on either environmental concentrations, pro-
jected concentrations, or toxicity.

PCBs: Bringing Good Things to Life?

PCBs are legacy chemicals. They provide one of the first examples of
the environmental consequences of releasing synthetic chemicals—
rarely, if ever, naturally produced on Earth—on a global scale. In the
United States, industrial production began in the late 1920s, and the
synthetic organochlorine chemicals soon found use in a number of
products, from paint to caulking to microscope oil, but their primary
use was as a cooling fluid in electrical transformers. The 209 different
structures (referred to as congeners) that make up commercial PCB
mixtures have two basic attributes in common: two six-carbon rings
(benzene rings), which rotate around a central bond depending on the
degree of chlorination; and chlorine molecules. The amount of chlorination and the shape assumed by the biphenyls because of chlorination determine the chemical nature of PCBs. The degree and position
of chlorines determine the viscosity of PCB oil (and therefore their
utility), the susceptibility of PCBs to be metabolized and degraded by
bacteria in the environment, and which PCBs will be detoxified in the
body and which will bind to and activate biological receptors. Those
PCBs resistant to environmental degradation and detoxification in living things are notoriously persistent and tend to be highly bioaccumu-
lative. Several PCB congeners, particularly those most resembling

dioxin in size, shape, and chemistry, bind with the aryl hydrocarbon receptor (discussed in chapter 7), albeit with different affinities for the receptor. These qualities have allowed toxicologists to develop structure-activity relationships for PCBs, which are useful for predicting toxicity across related species (reproductive and developmental toxicity, for example).[1] Congeners that do not interact with the receptor have their own toxicity profiles, particularly neurotoxicity.[2]

In 1976 the United States banned the manufacture and use of PCBs, with a more complete phaseout in the years that followed. PCBs were banned in part because of concern about their toxicity, but the finding that, like DDTs, they are highly persistent in the environment and readily accumulate in living things also contributed to their commercial demise. With more than 1.3 billion tons of PCBs produced worldwide from 1930 through their ban in the 1970s (or as late as 1990s, depending on the country), and an estimated 1.3%–12.4% still available for global transport through the year 2100, PCBs will be a part of life for years to come as they are "reemitted" from environmental reservoirs.[3] PCBs and similar chemicals that are currently locked away in glacial ice will likely be released and redistributed around the world as a result of climate change.[4] If there is one thing we have learned from PCBs, it is that simply banning a chemical does not make it go away.

While chapter 9 focused on the role of high concentrations of PCB on selection for resistant killifish and tomcod, PCBs are also thought to be partially responsible for the extirpation of local populations of mink and other susceptible species. Further, we know little about the long-term effects of persistent exposures, particularly in species that accumulate these chemicals through their diet. Species that accumulate PCBs also tend to accumulate other long-lasting, fat-loving chemicals, including PBDEs, DDTs, and methyl mercury. Some of these chemicals act similarly to PCBS and are detoxified through a similar process, while others interact with different regions of the defense network. As discussed in both chapters 1 and 10, we have spent nearly a century focused on the effects of individual chemicals. Today, we are fully aware that the world and life itself are far more complicated. Chemical mixtures may behave very differently than do single chemicals. Understanding how chemicals interact with biological systems (and in conjunction with other toxicants) could allow us to prevent harm at the outset of their production—and to avoid past mistakes like PCBs.

Mercury: It's a Mad, Mad, Mad, Mad World

Centuries-old anecdotes about "mad hatters" suggest that mercury, used in the felting process for hats, may have caused neurotoxicity in hatmakers. Sadly, the reproductive toxicity and neurotoxicity of the metal became better known because of several high-profile incidences of widespread exposure, including contamination of local fish and villagers following the release of mercury into Minamata Bay, Japan, by the Chisso Corporation.[5] Now we recognize that mercury can cause subtle damage, even in small concentrations.[6] It is one of the few *naturally* occurring chemical contaminants of importance today, and has long affected life—seldom, if ever, positively.[7] While some species, particularly bacteria, can adapt and survive toxic concentrations of mercury, others remain exquisitely sensitive.[8] Yet, despite all that we know of its toxicity, mercury contamination has been difficult to curtail. This is, in large part, because mercury is released when we burn coal and other fossil fuels.

The mercury we know as quicksilver, or liquid metal, is the least reactive, zero-oxidation state of the metal. As a vapor, this form is stable in the atmosphere, allowing for long distance transit—an important characteristic for mercury distribution around the globe. The loss of one electron leads to the mercurous form of the metal, while the loss of two electrons results in the mercuric ion, or divalent mercury. All forms are toxic, but methyl mercury—the result of a single mercuric ion combining with a single methyl group—is the most toxicologically important.[9] Divalent mercury is especially toxic because of its affinity for sulfur-containing thiol groups. Thiols are common in living things, particularly in the form of the amino acid cysteine or in molecules like glutathione, an enzyme important for detoxification. In fact, thiols are also known as "mercaptans" because of their penchant for "capturing mercury," a characteristic that makes them particularly vulnerable when the mercury rises. These relatively nonspecific combinations of metal and biological material are, in part, what make metals like mercury toxic.

Because of our reliance on resources once tucked away in the earth's crust, the amount of mercury available to living things has increased three to five times over naturally existing sources.[10] One study of bird eggshells from Guangjin Island in the South China Sea confirms both life's long history with mercury and the changes in metal availability associated with human activity.[11] Ancient eggshells from as far back as the fourteenth century contained measurable amounts of the metal. Yet the sudden rise in available mercury beginning in the early 1800s, accelerating in the last decade or so, provides a striking image: mercury concentrations in eggs from the modern era are up to ten times greater than in those from preindustrial times. Smaller spikes in mercury concentrations over the seven-hundred-year period of study may indicate volcanic eruptions or other natural releases of mercury, and at least in the 1600s, the beginning of heavy mining activity. The source of contamination to the island, as with so many globally distributed toxicants, is thought to be atmospheric transport from industrial regions. While *natural* sources release roughly 500 megagrams (1,000 kilograms, or one metric ton) each year, human activity is estimated to add an additional 2,000–4,000 megagrams per year and is on the rise.[12] Of mercury's fate in this world, concludes Noelle Selin in her review on the "Global Biogeochemical Cycling of Mercury":

> Mercury emitted to the atmosphere will remain in the atmosphere-ocean-terrestrial system for 3000 years before returning to the sediments. In the oceans, mercury levels have not yet reached steady state with respect to current levels of deposition. This means that if anthropogenic emissions continue at their present level, ocean concentrations in many ocean basins will increase in the future. Sulfate-reducing bacteria convert mercury into the toxic form of methylmercury. This process is affected by factors such as the sulfur cycle, ecosystem pH, and the presence of organic matter. Mercury has been regulated since the 1950s in industrialized countries and internationally since the 1970s, although global transport of mercury continues to be of concern, especially in the Arctic ecosystem. Bioaccumulation of mercury in the Arctic contaminates wildlife and traditional food sources.[13]

In other words, there's nothing "mercurial" or unpredictable about this silvery substance. We can safely say that increased concentrations of mercury will persist for centuries to come. Selin suggests that to reduce the amount of available mercury, we must do more than curtail "new" emissions. We must find ways to rein in the mercury released over hundreds of years of industrial activity that has already settled into soils, sediments, peatlands, and glaciers, *or* to prevent its conversion into readily available methylmercury. Life has a long history of mercury exposure, yet exposure today is not what it was millions of years ago. The evolution of complex food webs and top predators has resulted in bioaccumulation. The development of complex neural pathways and reproductive strategies may provide mercury with additional targets in some species—changes, perhaps, leading to greater susceptibility to this ancient toxicant.

Chlorofluorocarbons: Going, Going, Gone?

Chlorofluorocarbons (CFC) are chemicals that have affected life worldwide, albeit indirectly. A burgeoning human population, ironically seeking relief from the sun's heat, unleashed these fluorinated hydrocarbons. They are responsible for reductions in the earth's protective ozone layer, which (as discussed in chapter 2) have allowed greater penetration of the sun's ultraviolet light. First produced on an industrial scale in the 1930s as a refrigerant, these chemicals readily absorb heat energy as they transition from liquid to gas (and back). Like so many other chemical produced in the past century, CFCs resist degradation (at least under some circumstances) and may exist in the environment for upward of one hundred years. By the 1970s and 1980s, hundreds of thousands of pounds of CFCs were produced annually, primarily in three forms, referred to as CFC-11, -12, and -13. The *total production* of CFCs eventually peaked in the millions of metric tons, the large majority of which were eventually released to the environment.[14]

There was initially little concern about the release of CFCs because they are relatively nontoxic, nonflammable, and noncarcinogenic. Yet they are a prime contributor to stratospheric ozone depletion, which directly increases UVB exposure—and, in turn, UVB-induced DNA mutation and DNA repair. In some species, UVB is associated with increased skin cancer rates and other adverse effects. The 1987 UN Montreal Protocol on Substances That Deplete the Ozone Layer prompted a worldwide commitment to reduce the use and release of CFCs over several decades. The protocol's goal was to cut emissions in half by 2000 and to phase out CFC use by all countries, including developing nations, by 2010 at the latest.[15] One of the lingering holdouts or "essential" uses (and one that I've experienced personally) was as a propellant in asthma inhalers. This use accounted for more than a thousand tons per day but is now discontinued in the United States and other countries.[16]

Despite our best efforts, lingering CFCs will continue to influence the earth's ozone for years to come. Reporting on the status of the ozone hole in 2010, the United Nations World Meteorological Organization released the following statement:

> Depletion of the ozone layer—the shield that protects life on Earth from harmful levels of ultraviolet rays—has reached an unprecedented level over the Arctic this spring because of the continuing presence of the ozone-depleting substances in the atmosphere and a very cold winter in the stratosphere. The stratosphere is the second major layer of the Earth's atmosphere, just above the troposphere.
>
> The record loss is despite an international agreement which has been very successful in cutting production and consumption of ozone destroying chemicals. Because of the long atmospheric lifetimes of these compounds it will take several decades before their concentrations are back down to pre-1980 levels, the target agreed [to] in the Montreal Protocol on Substances That Deplete the Ozone Layer.[17]

Here again, we do not know how or *if* these changes, which work at the level of DNA, will affect the evolutionary trajectory of living things, from plankton to plants to mammals.

Endocrine Disrupters: The Hormonal Frontier

Organochlorines, mercury, and increased UVB represent only a small portion of human-induced changes or additions to the earth's chemistry. Whether synthetic or of natural origins, each contaminant interacts with highly conserved biological systems, including receptors, enzymes, and DNA. These chemicals are unlike those essential for life. So perhaps it is to be expected that life, defending itself for more than three billion years, developed *some* capacity to thwart both the known and the unknown. Yet how might living things respond to chemicals that behave like the physiologically essential chemicals that evolved to relay messages within or between cells? What happens when our systems have difficulty distinguishing essential chemical from contaminant? What are the effects of chemicals like environmental estrogens, antiandrogens, and thyroid hormones, which we have added to the earth's environment in unprecedented amounts? More important, how might populations respond to chemicals that interfere with reproductive success? Given the importance of reproduction, should we expect that defenses have evolved to protect life from an overabundance of estrogenic chemicals, or chemicals that block hormone activity? Could resistance occur through contemporary evolution, and what might that look like? How well can life defend against what only appear to be its own biochemicals? As one of the youngest disciplines within toxicology, the field of endocrine disrupters is moving fast, yet still there are many unknowns.

In his review "Environmental Signaling: What Embryos and Evolution Teach Us about Endocrine Disrupting Chemicals," the reproductive toxicologist John McLaughlin writes:

In environmental endocrine science, we have made a series of observations that, at first, seemed unconnected. However, now, as the observations start to establish a pattern, we can begin to discern the linkages between them. In the last 20 [years] we have discovered the intrinsic biological signaling properties of numerous synthetic environmental chemicals. We are also beginning to learn about the complex network of signaling molecules that facilitate information flow in the communication system of ecological life. In the same time period, cell and molecular biology has elucidated many of the signaling molecules necessary for intra- and intercellular communications. The similarities between the signaling strategies adopted by the internal and external world are probably more than coincidental if the evolution of the signaling systems followed, in any way, the convergent pathways suggested in this review.[18]

Of the endocrine system's relationship with external signals, continues McLaughlin:

Environmental signals are chemical messenger molecules functioning in a communication network linking numerous species.

One may speculate that the functional aspects of this more globally distributed network might have provided a framework or blueprint to build the internal communication networks of animals, which we call their endocrine systems. As such, similarities in response to such signals in some cases should not be unexpected. Indeed, a central strategy for all life forms is the transmission of important characteristics to their offspring.[19]

In chapter 7 of this book, we explored the highly conserved nature of these signaling systems, and in Chapter 8, the networked nature of response. Yet we are only just beginning to understand the short- and long-term consequences of interfering with these systems through large-scale introduction of endocrine-disrupting chemicals. McLaughlin concludes, "From an environmental stewardship perspective, the evolving concept of environmental signals can provide insights with which to address the impact of hormonally active chemicals on humans and the ecosystems that they share with other species. Disruption of this apparently broad communication system has the potential for global change that transcends the endocrine system."[20]

Ten years later, the list of chemicals that interact with the endocrine system continues to grow. Both the near-term consequences for individuals exposed to large amounts of endocrine-disrupting chemicals and the long-term consequences of chronic exposures to small amounts remain unclear. Yet the situation is ominous, given these chemicals' potential to disrupt reproduction, the most basic driver of fitness and therefore survival. The issue has received a great deal of attention lately with the advent of the Environmental Protection Agency's Endocrine Disruptor Screening Program and a slew of books, panels, and review articles.[21] The evolutionary history of receptors and chemical messengers could no doubt provide novel insights into ligand promiscuity, or lack thereof, in receptors—and perhaps lead to the development of chemicals less likely to take the place of a body's natural ligands.

Nanoparticles: Too Small to Ignore

Like endocrine disrupters, this last example is not about any single chemical, but rather a complex group of chemicals with one common characteristic: their exceedingly small size. Engineered nanoparticles are the fruits of nanotechnology, a development hailed by many as the "next" industrial revolution.[22] Though not yet a household world, nanomaterials are entering the consumer stream at a rapidly increasing rate. In *EPA and Nanotechnology: Oversight for the 21st Century*, Dr. J. Clarence (Terry) Davies writes, "In a few decades, almost every aspect of our existence is likely to be changed for the better by nano. However, if the potential for good is to be realized, society must also face nano's potential for harm."[23] In other words, the next industrial revolution is here. Are toxicologists, chemists, and regulators prepared to protect society and the environment from that potential harm?

Nanoparticles are both the newest additions to our chemical culture and, like mercury and UVB, also among the oldest. A nanoscale chemical, or nanoparticle, encompasses particulate chemical entities with dimensions less than one hundred nanometers (a billionth of a meter). Nanoparticles, or ultrafine particles, have existed in our atmosphere ever since there were fires, volcanoes, and sea spray to produce them. Viruses, made up of bits of RNA and DNA, are nano-sized. And ever since humans began burning coal and other fossil fuels on a large scale, combustion has added greatly to atmospheric nanoparticles. And now there is a chemical revolution centered on the intentional production of nano-sized particles, from single-walled carbon nanotubes to metal-based quantum dots, and composite nanomaterials containing DNA or other biomolecules. When reduced to a nanoscale, many chemicals exhibit physical and chemical properties that are different from their large-scale counterparts, essentially providing an opportunity for a whole new class of industrial chemical products. The nano-formulation of titanium dioxide, a chemical often used as a whitener, scatters very little visible light, resulting in a transparent but still effective sunscreen. Its wide use, combined with concerns about the adequacy of current techniques to evaluate toxicity (discussed below), prompted the EPA to use nanotitanium to develop its first case study of a nanomaterial. The objective was "to determine what is known and what needs to be known about selected nanomaterials as part of a process to identify and prioritize research to inform future assessments of the potential ecological and health implications of these materials."[24] Understanding how nanomaterials might behave inside living cells or in the environment will require in many cases new approaches to chemical evaluation. The evolutionary history of life's relationship with naturally occurring nanoparticles may provide insights into how engineered particles might behave, move through a body, or be absorbed by a digestive system.

Among those who first envisioned the technological potential of the very small was the Nobel Prize–winning physicist Richard Feynman. Toward the end of a 1959 lecture, Feynman offered a $1,000 reward to anyone who could figure out how produce a pinhead-sized version of the entire *Encyclopedia Britannica* (i.e., roughly one million nanometers across).[25] Though Feynman expected rapid progress, the technological breakthrough (and a legitimate claim for the reward) did not come for another thirty years. But since then, the production of nanoparticles has blossomed. Nanotechnology is now a billion-dollar field. Its potential environmental and health applications include more effective drug delivery, solar-derived power, reduced use of industrial chemicals, and improved environmental cleanup and decontamination.

Toxicologists, chemists, and regulators have a golden opportunity to learn from the past thirty years of chemical testing, management, and disasters and begin the era of nanotoxicology with an eye toward avoiding past mistakes. And to a certain extent, this is happening. Unlike industrial practices of the past, nanotechnology is unfolding under the scrutiny of government, public, and private organizations—in an era when reports, commentary, and scientific papers are more available than ever before to professionals and the public alike through the Internet.[26] Yet despite best efforts, fields like toxicology—particularly applied and regulatory toxicology—can take years to implement changes and new techniques. (For example, it has taken more than a decade to incorporate the testing of endocrine disruptors, despite the sense of urgency associated with it.)

Although toxicology sits on the cusp of major change, it is not quite ready for nanoparticles. As a cross between particulate and chemical toxicants, nanoparticles behave differently from their larger chemical counterparts. This is a large part of their allure to industry: nanoparticles open the door to a whole new world of chemical possibility. Yet these differences make challenging even some of the most basic procedures, like dosing, measurement of exposure, or distribution throughout a body. Commenting on the difficulty, a working group convened by the International Life Sciences Institute Research Foundation wrote in 2005, "There is a strong likelihood that biological activity of nanoparticles will depend on physicochemical parameters not routinely considered in toxicity screening studies."[27] In other words, today's techniques are inadequate for what are rapidly becoming today's chemicals. Different physicochemical parameters may also affect the behavior of particles in media typically used in preparing for traditional toxicity testing, the ability of researchers to adequately evaluate exposure concentrations, and particle behavior in the body. With more than one thousand "nanotech-enabled" consumer products already on the market—up from just two hundred back in 2006—the window for evaluating the behavior and toxicity of nanoparticles in both humans and in the environment, before their large-scale use and release, is closing.

The current body of literature on nanoparticles reveals the priorities and the rapid growth of nanoparticle research. A few years ago, an FDA task force reported that in the 1990s nanoparticle-related articles numbered only in the thousands, with two hundred patents worldwide. By 2002, publications numbered over 22,000 and patents, 1,900—a tenfold increase in ten years. But there is hope, as the nanotoxicology literature and research base grows. A broad literature search for "nano" and "toxic" resulted in three articles published in 1990 or before, 694 articles by 2002, and well over 9,000 articles by 2011. But this rapid expansion in research does not necessarily imply a coordination of techniques or data reporting. In fact, such rapid expansion might indicate haphazard approaches, as health and environmental scientists try to keep up with nanotechnology.

There are few commonalities when it comes to the chemical composition of nanoparticles. Yet most, perhaps all, share one characteristic that sets them apart from their larger counterparts: increased surface area. Typically, when something is produced in small form, its surface area increases. Consider peeling a pound of Granny Smith apples, for example, versus a pound of crab apples. Which is the more onerous task? More surface area, more skin on the smaller crab apples. It is the same with particulates: as they get smaller, they reveal more surface area and more individual atoms. Though the mass of a microgram dose of nanotitanium and titanium is the same, the nanotitanium has more surface area with which to react, and more reactive particles tend to be more toxic. As a result, one mechanism of toxicity shared by many nanoparticles is the generation of reactive oxygen species (ROS) and the toxicity associated with the stress response pathways triggered by ROS.[28] In addition to ROS, some nanoparticles may interact with cellular receptors, altering the signaling pathways initiated or sustained by activation of these receptors and perhaps inducing apoptosis, chromosomal instability, inflammation, or proliferation. Or nanoparticles may interfere with mitochondrial function, which can lead to apoptosis and inflammation. Or, more insidiously, some may interfere with protein folding.[29]

Those responses noted above and discussed throughout this book focus on biochemical or genomic reactions to chemicals. But of course, in complex life they all function within a larger context: that is, the whole organism, which has evolved many different layers and levels of protection and response. And while we have a fair understanding of how conventional chemicals move through a body, or a cell, this is not yet true for nanoparticles. We know, for example, which chemicals will easily cross a cell membrane or enter through specific pores and channels, and which will not. In more complex organisms, nutrients and environmental chemicals have only a few "portals of entry": the digestive tract or the pulmonary system. At such entry points, membranes and organs tend toward a greater concentration of defensive features, whether regions with higher concentrations of cytochrome P450s in the liver, patches of immune cells in the gut, or long passageways, cilia, and mucus-coated cells in the lung. Yet when it comes to the ability of nanoparticles to cross barriers, or be filtered out, we know very little. Do these portals regularly allow entry to nanosized chemicals? Will the pores and the gaps within and between our cells allow access to nanoparticles? How will the immune system respond? Some engineered particles are *designed* to avoid cellular exclusion and defensive devices.

While nanoparticles are nothing new to life, *engineered* nanoparticles present old chemicals in new packaging. These innovations come at a time when toxicology is maturing as a science. Advances in genomics and technology now provide us with greater understanding of individual genes and the networked nature of life's response to toxic chemicals, and a far greater capacity to trace the evolution of genes, enzymes, and receptors. These developments are changing how we evaluate and determine chemical toxicity. Toxicology is headed toward a paradigm shift. If that shift truly leads to improved knowledge, toxicology may help stimulate the nanotechnology revolution, while reining in the collateral damage to life and the environment so typical of past revolutions.

Acknowledgments

1. A. Rico, "Chemo-Defense System," *Comptes Rendus de l'Académie des Sciences Series III: Sciences de la Vie* 324 (2000): 105.

Chapter 1

1. Reviewed in A. Hendry et al., "Evolutionary Principles and Their Practical Application," *Evolutionary Applications* 4 (2011): 159–83; S. Palumbi, "Better Evolution through Chemistry: Rapid Evolution Driven by Human Changes to the Chemical Environment," *Chemical Evolution II: From the Origins of Life to Modern Society*, ACS Symposium Series 1025 (2009): 333–43; M. Coutellec and C. Barata, "An Introduction to Evolutionary Processes in Ecotoxicology," *Ecotoxicology* 20 (2011): 493–96.

2. A special issue of the journal *Ecotoxicology* highlights the importance of considering evolutionary processes in ecological risk assessment, particularly in relation to rapid evolution in contaminant-exposed populations. For an overview, see Coutellec and Barata, "Introduction to Evolutionary Processes." For discussion of the influence of contaminants on genetic diversity, see J. W. Bickham, "The Four Cornerstones of Evolutionary Toxicology," *Ecotoxicology* 20 (2011): 497–502; and J. Bickham and M. Smolen, "Somatic and Heritable Effects of Environmental Genotoxins and the Emergence of Evolutionary Toxicology," *Environmental Health Perspectives* 102 (suppl. 2) (1994): 25–28.

3. T. Dobzhansky, "Nothing in Biology Makes Sense except in the Light of Evolution," *American Biology Teacher* 35 (1973): 125–29.

4. J. Bull and H. Wichman, "A Revolution in Evolution," *Science* 281 (1998): 1959.

5. F. Gonzalez and D. Nebert, "Evolution of the P450 Gene Superfamily," *Trends in Genetics* 6 (1990): 182–86.

6. C. Theodorakis, "Establishing Causality between Population Genetic Alterations and Environmental Contamination in Aquatic Organisms," *Human and Ecological Risk Assessment* 9 (2003): 37–58; A. Morgan, P. Kille, and S. Struzenbaum, "Microevolution and Ecotoxicology of Metals in Invertebrates," *Environmental Science and Technology* 41 (2007): 1085–96; Bickham

and Smolen, "Somatic and Heritable Effects of Environmental Genotoxins";
T. Moyer et al., "Warfarin Sensitivity Genotyping," *Mayo Clinic Proceedings* 84
(2009): 1079–94. See also "Warfarin Effectiveness Study," March 2010,
Mayo Clinic, accessed June 2011, http://www.mayomedicallaboratories.com
/tests/warfarin/index.html; L. Costa and D. Eaton, eds., *Gene-Environment
Interactions: Fundamentals of Ecogenetics* (Hoboken, NJ: Wiley, 2006).

7. The toxicologist Andre Rico is one of the first to suggest that life
evolved a coordinated system for "chemo-defense." See "Chemo-Defense System," *Comptes Rendus de l'Académie des Sciences Series III: Sciences de la Vie* 324
(2000): 97–106.

8. N. Robins and J. Evans, "Why Physicians Must Understand Evolution," *Current Opinion in Pediatrics* 21 (2009): 699.

9. "Ongoing Human Evolution Could Explain Recent Rise in Certain
Disorders," *Science Daily*, January 11, 2011, accessed July 2011, http://www
.sciencedaily.com/releases/2010/01/100111102538.htm. For further reading, see P. Gluckman, A. Beedle, and M. Hanson, eds., *Principles of Evolutionary Medicine* (Oxford: Oxford University Press, 2010).

10. R. M. Nesse, S. C. Stearns, and G. S. Omenn, "Why Medicine Needs
Evolution," *Science* 311 (2006): 1071

11. H. Crone, *Paracelsus: The Man Who Defied Medicine* (Melbourne: Al-
barello Press, 2004), 121.

12. L. Gerber, G. Williams, and S. Gray, "The Nutrient-Toxin Dosage
Continuum in Human Evolution and Modern Health," *Quarterly Review of
Biology* 74 (1999): 273–89.

13. R. Williams and J. Frausto da Silva, *The Chemistry of Evolution* (New
York: Elsevier, 2006).

14. F. Egami, "Origin and Early Evolution of Transition Element Enzymes," *Journal of Biochemistry* 77 (1974): 1165–69; T. Clarkson, "Health Effects of Metals: A Role for Evolution?," *Environmental Health Perspectives* 103
(suppl. 1) (1995): 9–12.

15. A. Summers, "Damage Control: Regulating Defenses against Toxic
Metals and Metalloids," *Current Opinion in Microbiology* 12 (2009): 138–44.

16. R. Ruch, "Intercellular Communication, Homeostasis, and Toxicology," *Toxicological Sciences* 68 (2002): 265–66.

17. Clarkson, "Health Effects of Metals."

18. J. Lewis, "Lead Poisoning: A Historical Perspective," *EPA Journal*,
May 1985, accessed June 2011, http://www.epa.gov/aboutepa/history/topics
/perspect/lead.html.

19. For a modern example of the effects of large-scale lead contamination, see G. Markowitz and D. Rosner, *Deceit and Denial: The Deadly Politics of
Industrial Pollution* (Berkeley: University of California Press, 2002).

20. Ruch, "Intercellular Communication, Homeostasis, and Toxicology";

for details about hormesis, see E. Calabrese, "Converging Concepts: Adaptive Response, Reconditioning, and the Yerkes-Dodson Law Are Manifestations of Hormesis," *Ageing Research Reviews* 7 (2008): 8–20; E. Calabrese and L. Baldwin, "Chemical Hormesis: Its Historical Foundations as a Biological Hypothesis," *Human and Experimental Toxicology* 19 (2000): 2–31; P. Parsons, "The Hormetic Zone: An Ecological and Evolutionary Perspective Based upon Habitat Characteristics and Fitness Selection," *Quarterly Review of Biology* 76 (2001): 459–67.

21. Ruch, "Intercellular Communication, Homeostasis, and Toxicology."

22. See T. Colborn and C. Clement, eds., *Chemically-Induced Alternations in Sexual and Functional Development: The Wildlife/Human Connection* (Princeton, NJ: Princeton Scientific, 1992).

23. T. Hartung, "Toxicology for the Twenty-First Century," *Nature* 460 (2009): 208. The National Research Council's report on toxicology for the twenty-first century suggests that he is not alone in calling for an overhaul, since it includes as one of its goals the "develop[ment of] a more robust scientific basis for assessing health effects of environmental agents." National Research Council, *Toxicity Testing in the 21st Century* (Washington, DC: National Academies Press, 2007), accessed June 2011, http://www.nap.edu /openbook.php?record_id=11970&page=1.

24. E. Koonin, "Darwinian Evolution in the Light of Genomics," *Nucleic Acids Research* 37 (2009): 1011–34.

25. M. Hahn, "Aryl Hydrocarbon Receptors: Diversity and Evolution," *Chemico-Biological Interactions* 141 (2002): 131–60; J. Goldstone, "Environmental Sensing and Response Genes in Cnidaria: The Chemical Defensome in the Sea Anemone *Nematostella vectensis*," *Cell Biology and Toxicology* 24 (2008): 483–502; B. May and P. Dennis, "Evolution and Regulation of the Gene Encoding Superoxide Dismutase from the Archaebacterium *Halobacterium cutirubrum*," *Journal of Biological Chemistry* 264 (1989): 12253–58; C. Andreini, I. Bertini, and A. Rosato, "Metalloproteomes: A Bioinformatic Approach," *Accounts of Chemical Research* 42 (2009): 1471–79; C. Dupont et al., "Modern Proteomes Contain Putative Imprints of Ancient Shifts in Trace Metal Geochemistry," *Proceedings of the National Academy of Sciences of the United States of America* 103 (2006): 17822–27.

26. D. Conover and S. Munch, "Sustaining Fisheries Yields over Evolutionary Time Scales," *Science* 297 (2002): 94–96; C. Darimont et al., "Human Predators Outpace Other Agents of Trait Change in the Wild," *Proceedings of the National Academy of Sciences of the United States of America* 106 (2009): 952–54.

27. Reviewed in A. Hendry and M. Kinnison, "The Pace of Modern Life: Measuring Rates of Contemporary Microevolution," *Evolution* 53 (1999): 1637–53; Bickham and Smolen, "Somatic and Heritable Effects of

Environmental Genotoxins"; Theodorakis, "Establishing Causality between Population Genetic Alterations and Environmental Contamination in Aquatic Organisms."

28. Hendry and Kinnison, "Pace of Modern Life"; Dobzhansky, "Nothing in Biology Makes Sense except in the Light of Evolution."

29. See, for example, J. W. Bickham et al., "Effects of Chemical Contaminants on Genetic Diversity in Natural Populations: Implications for Biomonitoring and Ecotoxicology," *Mutation Research* 463 (2000): 33–51; G. Maes et al., "The Catadromous European Eel *Anguilla anguilla* (L.) as a Model for Freshwater Evolutionary Ecotoxicology: Relationship between Heavy Metal Bioaccumulation, Condition, and Genetic Variability," *Aquatic Toxicology* 73 (2005): 99–114; V. Bol'shakov and T. Moiseenko, "Anthropogenic Evolution of Animals: Facts and Their Interpretation," *Russian Journal of Ecology* 40 (2009): 305–13.

30. Conover and Munch, "Sustaining Fisheries Yields over Evolutionary Time Scales."

31. Hendry and Kinnison, "Pace of Modern Life."

32. S. Carroll, *Endless Forms Most Beautiful* (New York: Norton, 2005), 39.

Chapter 2

1. S. Miller and H. Urey, "Organic Compound Synthesis on the Primitive Earth," *Science* 130 (1959): 245–51; A. Johnson et al., "The Miller Volcanic Spark Discharge Experiment," *Science* 322 (2008): 404. See also D. Fox, "Primordial Soup's On: Scientists Repeat Evolution's Most Famous Experiment," *Scientific American*, March 28, 2007, accessed June 2011, http://www.scientificamerican.com/article.cfm?id=primordial-soup-urey-miller-evolution-experiment-repeated.

2. "Did Comets Contain Key Ingredients for Life on Earth?," *Science Daily*, April 29, 2009, accessed April 2010, http://www.sciencedaily.com/releases/2009/04/090428144126.htm; and "Sweet Meteorites," *NASA Science News*, December 20, 2001, accessed June 2011, http://science.nasa.gov/science-news/science-at-nasa/2001/ast20dec_1/.

3. N. Lane, J. Allen, and W. Martin, "How Did LUCA Make a Living? Chemiosmosis in the Origin of Life," *BioEssays* 32 (2011): 271–80.

4. A. Ricardo and J. Szostak, "Life on Earth," *Scientific American* 301 (2009): 54–61.

5. B. Diffey, "Solar Ultraviolet Radiation Effects on Biological Systems," *Center for International Earth Science Information Network*, July 30, 1990, accessed July 2011, http://www.ciesin.columbia.edu/docs/001-503/001-503.html; "Ultraviolet and Visible Light Spectroscopy," *University of California, Davis, ChemWiki*, n.d., accessed July 2011, http://chemwiki.ucdavis

.edu/Organic_Chemistry/Organic_Chemistry_With_a_Biological_Emphasis /Chapter__4:_Structure_Determination_I/Section_4.3:_Ultraviolet_and _visible_spectroscopy.

6. United States Federal Food and Drug Administration General and Plastic Surgery Devices Panel of the Medical Devices Advisory Committee, "Executive Summary," FDA.gov, March 25, 2010, accessed July 2011, http://www.fda.gov/downloads/AdvisoryCommittees/CommitteesMeeting Materials/MedicalDevices/MedicalDevicesAdvisoryCommittee/Generaland PlasticSurgeryDevicesPanel/UCM205687.pdf.

7. For more about vitamin D, see N. Seppa, "The Power of D," *Science News* 180 (2011): 22–26; and Daniel D. Bikle, "Vitamin D: An Ancient Hormone," *Experimental Dermatology* 20 (2011): 7–13.

8. B. Diffey, "Solar Ultraviolet Radiation Effects on Biological Systems," *Physics in Medicine and Biology* 36 (1990): 299–328.

9. For a discussion of radiation and solar wind, see J. Tarduno et al., "Geodynamo, Solar Wind, and Magnetopause 3.4 to 3.45 Billion Years Ago," *Science* 327 (2010): 1238–40. See also J. Jardine, "Sunscreen for the Young Earth," *Science* 327 (2010): 1206–8.

10. For further reading on the origins of oxygen, see N. Lane, *Oxygen: The Molecule That Made the World* (Oxford: Oxford University Press, 2009).

11. Ibid., 16; C. Cockell and A. Blaustein, eds., *Ecosystems, Evolution, and Ultraviolet Radiation* (New York: Springer, 2001).

12. Lane, *Oxygen*.

13. Ibid., 72; C. Cockell, "Biological Effects of High Ultraviolet Radiation on Early Earth—A Theoretical Evaluation," *Journal of Theoretical Biology* 193 (1998): 717–19.

14. Cockell, "Biological Effects of High Ultraviolet Radiation on Early Earth"; Cockell and Blaustein, *Ecosystems, Evolution, and Ultraviolet Radiation*.

15. C. Cockell, "Ultraviolet Radiation and the Photobiology of Earth's Early Oceans," *Origins of Life and Evolution of the Biosphere* 30 (2000): 467–99.

16. Since most UVC is filtered out by the ozone layer, and UVA causes other damage (including the formation of reactive species, discussed in chapter 3), the primary damage cause by ambient UVR stems from UVB.

17. L. Essen and T. Klar, "Light-Driven DNA Repair by Photolyases," *Cellular and Molecular Life Sciences* 63 (2006): 1266–77; R. Sinha and D. Hader, "UV-Induced DNA Damage and Repair: A Review," *Photochemical and Photobiological Sciences* 1 (2002): 225–36.

18. Essen and Klar, "Light-Driven DNA Repair by Photolyases"; Sinha and Hader, "UV-Induced DNA Damage and Repair."

19. B. Grant, "Should Evolutionary Theory Evolve?," *The Scientist* 24 (2010): 24–30.

20. J. Lucas-Lledo and M. Lynch, "Evolution of Mutation Rates: Phylogenomic Analysis of the Photolyase/Cryptochrome Family," *Molecular Biology and Evolution* 26 (2009): 1143–53.

21. Interestingly RNA, the messenger for DNA, and thought by some to be the precursor to DNA, is far more resistant to UV damage.

22. E. Koonin, "Darwinian Evolution in Light of Genomics," *Nucleic Acids Research* 37 (2009): 1011–34. This article reviews the potential role of "purifying selection," which holds that the intensity of purifying selection is proportional to the effective population size. That is, purifying selection may be far more efficient in bacteria (where the majority of DNA encodes for proteins) than in multicellular animals (where the majority of DNA does not, and instead is considered "junk," until further analysis).

23. See J. Kimball, "Biology Pages," accessed July 2011, http://users.rcn.com/jkimball.ma.ultranet/BiologyPages/G/GenomeSizes.html.

24. See J. Sapp, *The New Foundations of Evolution: On the Tree of Life* (Oxford: Oxford University Press, 2009).

25. Reviewed in A. Poole, "My Name Is LUCA," *ActionBioscience*, September 2002, accessed July 2011, http://www.actionbioscience.org/newfrontiers/poolepaper.html; and Sapp, *New Foundations of Evolution*, 286–99. See also N. Glansdorff, Y. Xu, and B. Labedan, "The Last Universal Common Ancestor: Emergence, Constitution, and Genetic Legacy of an Elusive Forerunner," *Biology Direct* 3 (2008): 29.

26. See P. Forterre, "The Search for LUCA," *1997 Workshop Proceedings*, accessed July 2011, http://translate.google.com/translate?hl=en&sl=fr&u=http://www.-archbac.u-psud.fr/meetings/lestreilles/treilles_frm.html&ei=ZzHXS87dOIaglAetkriABA&sa=X&oi=translate&ct=result&resnum=2&ved=0CAwQ7gEwAQ&prev=/search%3Fq%3Dbaptizing%2BLUCA%2Bouzounis%26hl%3Den.

27. C. Ouzounis et al., "A Minimal Estimate for the Gene Content of the Last Universal Common Ancestor—Exobiology from a Terrestrial Perspective," *Research in Microbiology* 157 (2006): 57–68.

28. Ibid.; Glansdorff, Xu, and Labedan, "Last Universal Common Ancestor."

29. Sinha and Hader, "UV-Induced DNA Damage and Repair"; T. Kato et al., "Cloning of a Marsupial DNA Photolyase Gene and the Lack of Related Nucleotide Sequences in Placental Mammals," *Nucleic Acids Research* 22 (1994): 4119–24; C. Menck, "Shining a Light on Photolyases," *Nature Genetics* 32 (2002): 338–39.

30. S. Kumar and B. Hedges, "A Molecular Timescale for Vertebrate Evolution," *Nature* 392 (1998): 917–20.

31. Lucas-Lledo and Lynch, "Evolution of Mutation Rates."

32. P. O'Brien, "Catalytic Promiscuity and the Divergent Evolution of

DNA Repair Enzymes," *Chemical Reviews* 106 (2006): 720–52. According to O'Brien, humans do have closely related proteins that appear to be involved in another light-related activity: setting the circadian clock.

33. Reviewed in Koonin, "Darwinian Evolution in Light of Genomics."

34. Lucas-Lledó and Lynch, "Evolution of Mutation Rates"; M. Lynch, "The Origins of Eukaryotic Gene Architecture," *Molecular Biological Evolution* 23 (2006): 450–68.

35. Lucas-Lledó and Lynch, "Evolution of Mutation Rates," 1149.

36. T. Lindahl and R. Wood, "Quality Control by DNA Repair," *Science* 286 (1999): 1897–1905. Note that a number of repair mechanisms for DNA damage in general exist in most species, including humans; however, discussion of DNA repair in general is well beyond the scope of this chapter. See also Sinha and Hader, "UV-Induced DNA Damage and Repair."

37. Lindahl and Wood, "Quality Control by DNA Repair"; Sinha and Hader, "UV-Induced DNA Damage and Repair."

38. L. Briemer, "The Real Kaposi's Sarcoma," *Nature Medicine* 2 (1996): 131.

39. J. DiRuggiero et al., "DNA Repair Systems in Archaea: Mementos from the Last Universal Common Ancestor?," *Journal of Molecular Evolution* 49 (1999): 474–84; Ouzounis et al., "Minimal Estimate for the Gene Content of the Last Universal Common Ancestor."

40. T. Lindahl and T. Wood, "Quality Control by DNA Repair," *Science* 286 (1999): 1903.

41. P. O'Brien, "Catalytic Promiscuity and the Divergent Evolution of DNA Repair Enzymes," *Chemical Reviews* 106 (2006): 720–52.

42. A. Blaustein and L. Belden, "Amphibian Defenses against Ultraviolet-B Radiation," *Evolution and Development* 5 (2003): 89–97.

43. J. Farman, B. Gardiner, and J. Shanklin, "Large Losses of Total Ozone in Antarctica Reveal Seasonal ClOx/NOx Interaction," *Nature* 315 (1985): 207–10.

44. For a review, see S. Solomon, "The Hole Truth," *Nature* 427 (2004): 289–91; and F. Sherwood Rowland, "Stratospheric Ozone Depletion by Chlorofluorocarbons (Nobel Lecture)," *The Encyclopedia of Earth*, April 27, 2007, accessed July 2011, http://www.eoearth.org/article/Stratospheric _Ozone_Depletion_by_Chlorofluorocarbons_(Nobel_Lecture)#Ozone _Losses_in_the_Northern_Temperate_Zone.

45. United Nations Environment Programme, "Basic Facts and Data on the Science and Politics of Ozone Depletion," September 2008, accessed August 2011, http://ozone.unep.org/Events/ozone_day_2009/press _backgrounder.pdf; United Nations Environment Programme, "Environmental Effects of Ozone Depletion: 1991 Update," November 1991, accessed July 2011, http://www.ciesin.org/docs/011-558/011-558.html.

46. Unfortunately some of the more recent fluorinated replacements contribute to climate change. See UNEP, "Basic Facts and Data on the Science and Politics of Ozone Depletion."

47. S. Diaz et al., "Ozone and UV Radiation over Southern South America: Climatology and Anomalies," *Photochemistry and Photobiology* 82 (2006): 834–43.

48. Rowland, "Stratospheric Ozone Depletion by Chlorofluorocarbons."

49. S. Stuart et al., "Status and Trends of Amphibian Declines and Extinctions Worldwide," *Science* 306 (2004): 1783–86.

50. B. Bancroft, N. Baker, and A. Blaustein, "A Meta-Analysis of the Effects of Ultraviolet B Radiation and Its Synergistic Interactions with pH, Contaminants, and Disease on Amphibian Survival," *Conservation Biology* 22 (2008): 987–96.

51. Blaustein and Belden, "Amphibian Defenses against Ultraviolet-B Radiation"; Bancroft, Baker, and Blaustein, "Meta-Analysis of the Effects of Ultraviolet B Radiation"; A. Blaustein, "Amphibians in a Bad Light," *Natural History* 103 (1994): 32–37.

52. Blaustein, "Amphibians in a Bad Light."

53. O. Marquis et al., "Variation in Genotoxic Stress Tolerance among Frog Populations Exposed to the UV and Pollutant Gradients," *Aquatic Toxicology* 95 (2009): 152–61.

54. Ibid.

55. J. Klesecker, A. Blaustein, and L. Belden, "Complex Causes of Amphibian Population Declines," *Nature* 410 (2001): 681–84.

56. M. Stice and C. Briggs, "Immunization Is Ineffective at Preventing Infection and Mortality Due to the Amphibian Chytrid Fungus *Batrachochytrium dendrobatidis*," *Journal of Wildlife Diseases* 46 (2010): 70–77; E. Rosenblum et al., "Genome-Wide Transcriptional Response of *Silurana (Xenopus) tropicalis* to Infection with the Deadly Chytrid Fungus," *PLoS ONE* 4 (2008): e6494.

Chapter 3

1. "It's Elemental: The Periodic Table of the Elements," *Jefferson Lab*, n.d., accessed July 2011, http://education.jlab.org/itselemental/ele008.html.

2. R. Hazen, "Evolution of Minerals," *Scientific American* 302 (2010): 58–65.

3. While O_2 is common, atomic or single molecules of oxygen are nonexistent in the earth's atmosphere. Single oxygen molecules do exist in space (they are formed by solar UV—which causes O_2 to decompose—and are highly reactive), where it is highly corrosive to spacecraft. In fact, efforts to protect spacecraft inadvertently led to using atomic oxygen to clean old art by removing the buildup of organics, including soot. For more, see "Destructive

Power of Atomic Oxygen Used to Restore Art," *Glenn Research Center*, January 4, 2005, accessed July 2011, http://www.nasa.gov/centers/glenn/business/AtomicOxRestoration.html.

4. N. Lane, *Oxygen: The Molecule That Made the World* (Oxford: Oxford University Press, 2004), 120.

5. I. Fridovich, "Oxygen Toxicity: A Radical Explanation," *Journal of Experimental Biology* 201 (1998): 1203–9.

6. Lane, *Oxygen*, 121–22.

7. See Lane, *Oxygen*; Eric Chaisson, *The Life Era: Cosmic Selection and Conscious Evolution* (Bloomington, IN: iUniverse, 2000); R. Williams and J. Frausto da Silva, *The Chemistry of Evolution* (New York: Elsevier, 2006).

8. J. Kirschvink and R. Kopp, "Palaeoproterozoic Ice Houses and the Evolution of Oxygen-Mediating Enzymes: The Case for a Late Origin of Photosystem II," *Philosophical Transactions of the Royal Society, Series B: Biological Sciences* 363 (2008): 2756. Some like John Saul have also suggested that the sequestration of the gas in oxygen-rich molecules may also have provided some respite—however, such molecules, like collagen, are not widespread in unicellular life. See J. Saul, "Did Detoxification Processes Cause Complex Life to Emerge?," *Lethaia* 42 (2009): 179–84.

9. See "The Power of Earth and Beyond: Unique Approach for Splitting Water into Hydrogen and Oxygen," *Energy Daily*, April 7, 2009, accessed July 2011, http://www.energy-daily.com/reports/Unique_Approach_For_Splitting_Water_Into_Hydrogen_And_Oxygen_999.html.

10. While most of us equate photosynthesis with oxygen, in its earliest form, photosynthesis was the provenance of our anoxic oxygen–intolerant ancestors, who split molecules like hydrogen sulfide, or H_2S, rather than water as they gleaned energy from their surrounding environment. As they lived, respired, and died, they left behind laminated sedimentary structures recognized today as stromatolites, enabling scientists to push back the earliest time for photosynthetic life to somewhere around 3.5 billion years ago. For more, see S. Awramik, "Paleontology: Respect for Stromatolites," *Nature* 441 (2006): 700–701; M. Tice and D. R. Lowe, "Photosynthetic Microbial Mats in the 3,416-Myr-Old Ocean," *Nature* 431 (2004): 549–52.

11. J. Raymond and R. Blankenship, "The Origin of the Oxygen-Evolving Complex," *Coordination Chemistry Reviews* 252 (2008): 377–83; L. Björn and Govindjee, "The Evolution of Photosynthesis and Chloroplasts," *Current Science* 96 (2009): 1466–74. See also P. Preuss, "Spinach, or the Search for the Secret of Life as We Know It," *Berkeley Lab News Center*, July 14, 2004, accessed July 2011, http://newscenter.lbl.gov/feature-stories/2004/07/14/spinach-or-the-search-for-the-secret-of-life-as-we-know-it/; D. Des Marais, "When Did Photosynthesis Emerge on Earth?," *Science* 289 (2000): 1703–5.

12. For more on the evolution of photosynthesis see N. Lane, *Life Ascending: The Ten Great Innovations of Evolution* (New York: Norton, 2009).

13. H. Ohmoto et al., "Sulphur Isotope Evidence for an Oxic Archaean Atmosphere," *Nature* 442 (2006): 908–91; and see Lane, *Oxygen*, chaps. 3 and 4; A. Sessions et al., "The Continuing Puzzle of the Great Oxidation Event," *Current Biology* 19 (2009): R567–74.

14. James Lovelock, *Gaia: A New Look at Life on Earth* (Oxford: Oxford University Press, 1987), 31.

15. See Lane, *Oxygen*, 21.

16. J. Saul, Paris, e-mail communication, April 8, 2011.

17. T. Lyons and C. Reinhard, "Early Earth: Oxygen for Heavy-Metal Fans," *Nature* 461 (2009): 179–81. Other evidence for oxygen includes the nuclear reactors of Gabon and elsewhere; see L. Coogan and J. Cullen, "Did Natural Reactors Form as a Consequence of the Emergence of Oxygenic Photosynthesis during the Archean?," *Geological Society of America Today* 19 (2009): 4–10; P. Berardelli, "Did Ancient Earth Go Nuclear?," *ScienceNOW*, October 29, 2009, accessed July 2011, http://news.sciencemag.org/sciencenow/2009/10/29-01.html.

18. For discussion about oxygen evolution, see Kirschvink and Kopp, "Palaeoproterozoic Ice Houses and the Evolution of Oxygen-Mediating Enzymes"; A. Anbar et al., "A Whiff of Oxygen before the Great Oxidation Event?," *Science* 317 (2007): 1903–6.

19. Lane, *Oxygen*, 169

20. P. Falkowski, "Tracing Oxygen's Imprint on Earth's Metabolic Evolution," *Science* 311 (2006): 1724–67; Coogan and Cullen, "Did Natural Reactors Form as a Consequence of the Emergence of Oxygenic Photosynthesis during the Archean?"; Lane, *Oxygen*, chaps. 2 and 3.

21. E. Monosson, "Chemical Mixtures: Considering the Evolution of Toxicology and Chemical Assessment," *Environmental Health Perspectives* 113 (2004): 383–90.

22. K. Barbusiński, "Henry John Horstman Fenton—Short Biography and Brief History of Fenton Reagent Discovery," *Chemia, Dydaktyka, Ekologia, Metrologia* 14 (2009): NR1–2; A. Evens, J. Mehta, and L. Gordon, "Rust and Corrosion in Hematopoietic Stem Cell Transplantation: The Problem of Iron and Oxidative Stress," *Bone Marrow Transplantation* 34 (2004): 561–71.

23. I. Benzie, "Evolution of Antioxidant Defense Mechanisms," *Eur Jour Nutri* 39 (2000): 53–61.

24. Ibid.; Saul, "Did Detoxification Processes Cause Complex Life to Emerge?"

25. J. Massabuau, "Primitive, and Protective, Our Cellular Oxygenation Status?," *Mechanisms of Aging and Development* 124 (2003): 857–63.

26. R. Danovaro et al., "The First Metazoan Living in Permanently Anoxic Conditions," *Biomed Central Biology* 8 (2003): 30.

27. M. Di Giulio, "The Universal Ancestor and the Ancestors of Archaea and Bacteria Were Anaerobes Whereas the Ancestor of the Eukarya Do-

main Was an Aerobe," *European Society for Evolutionary Biology* 20 (2007): 543–48.

28. Lane, *Oxygen*, 168; J. Castresana and M. Seraste, "Evolution of Energetic Metabolism: The Respiration-Early Hypothesis," *Trends in Biological Sciences* 20 (1995): 443–48.

29. C. Ouzounis et al., "A Minimal Estimate for the Gene Content of the Last Universal Common Ancestor—Exobiology from a Terrestrial Perspective," *Research in Microbiology* 157 (2006): 57–68.

30. Those who suggest LUCA was more likely anaerobic include Jason Raymond and Daniel Segre. See J. Raymond and D. Segre, "The Effect of Oxygen on Biochemical Networks and the Evolution of Complex Life," *Science* 311 (2007): 1764–67; and Di Giulio, "The Universal Ancestor and the Ancestors of Archaea and Bacteria Were Anaerobes."

31. L. Ji et al., "Exercise-Induced Hormesis May Help Healthy Ageing," *Dose-Response* 8 (2010): 73–79; M. Nikolaidis and A. Jamurtas, "Blood as a Reactive Species Generator and Redox Status Regulator during Exercise," *Archives of Biochemistry and Biophysics* 490 (2009): 77–84; S. Sachdev and K. Davies, "Production, Detection, and Adaptive Responses to Free Radicals in Exercise," *Free Radical Biology and Medicine* 44 (2008): 215–23.

32. Lane, *Oxygen*, 140. See also D. Goodsell, "Catalase, September 2004 Molecule of the Month," *RCSB Protein Data Bank*, n.d., accessed July 2011, http://www.rcsb.org/pdb/static.do?p=education_discussion/molecule_of_the_month/pdb57_1.html.

33. E. Skaar, "A Precious Metal Heist," *Cell Host and Microbe* 5 (2009): 422–24.

34. B. Banerjee, V. Seth, and R. Ahmed, "Pesticide-Induced Oxidative Stress: Perspectives and Trends," *Reviews on Environmental Health* 16 (2001): 1–29; H. Andersson et al., "Plasma Antioxidant Responses and Oxidative Stress Following a Soccer Game in Elite Female Players," *Scandinavian Journal of Medicine and Science in Sports* 20 (2010): 600–608.

35. For example, see F. Tao, B. Gonzalez-Flecha, and L. Kobzik, "Reactive Oxygen Species in Pulmonary Inflammation by Ambient Particulates," *Free Radical Biology and Medicine* 35 (2003): 327–40.

36. For a discussion of past practices, see W. Silverman, "A Cautionary Tale about Supplemental Oxygen: The Albatross of Neonatal Medicine," *Pediatrics* 113 (2004): 394–96; L. Stern, "Oxygen Toxicity in Premature Infants," *Albrecht von Graefe's Archive for Clinical and Experimental Ophthalmology* 975 (1975): 71–76.

37. R. Jenkins, "Exercise and Oxidative Stress Methodology," *American Journal of Clinical Nutrition* 72 (suppl.) (2000): 670S–74S.

38. Andersson et al., "Plasma Antioxidant Responses and Oxidative Stress." During high activity, however, mitochondrial ROS production does not increase proportionally—reducing the potential for "massive" oxidative

damage, as noted in D. Costantini, "Oxidative Stress in Ecology and Evolution: Lessons from Avian Studies," *Ecology Letters* 11 (2008): 1238–51.

39. Z. Radak et al., "Exercise, Oxidative Stress, and Hormesis," *Ageing Research Reviews* 7 (2008): 34–42.

40. Ibid. For other reviews, see Ji et al., "Exercise-Induced Hormesis"; Andersson et al., "Plasma Antioxidant Responses and Oxidative Stress."

41. J. Scandalios, "Oxidative Stress: Molecular Perception and Transduction of Signals Triggering Antioxidant Gene Defenses," *Brazilian Journal of Medical and Biological Research* 38 (2005): 995–1014; Radak et al., "Exercise, Oxidative Stress, and Hormesis"; Ji et al., "Exercise-Induced Hormesis."

42. For exercise as an antioxidant, see M. Gomez-Cabrera and D. Vina, "Moderate Exercise Is an Antioxidant: Upregulation of Antioxidant Genes by Training," *Free Radical Biology and Medicine* 44 (2008): 126–31. By some estimates, tens of thousands of genes respond to ROS; see Scandalios, "Oxidative Stress."

43. R. Jenkins, "Exercise and Oxidative Stress Methodology"; Sachdev and Davies, "Production, Detection, and Adaptive Responses to Free Radicals in Exercise"; C. Scheele, S. Nielson, and B. Pederson, "ROS and Myokines Promote Muscle Adaptation to Exercise," *Trends in Endocrinology and Metabolism* 20 (2008): 95–99.

44. Costantini, "Oxidative Stress in Ecology."

45. K. McGraw et al., "The Ecological Significance of Antioxidants and Oxidative Stress: A Marriage between Mechanistic and Functional Perspectives," *Functional Ecology* 24 (2010): 947–49; Costantini, "Oxidative Stress in Ecology."

46. N. Metcalfe and C. Alonso-Alvarez, "Oxidative Stress as a Life-History Constraint: The Role of Reactive Oxygen Species in Shaping Phenotypes from Conception to Death," *Functional Ecology* 24 (2010): 984–96.

47. I. Chirino et al., "PM_{10} Impairs the Antioxidant Defense System and Exacerbates Oxidative Stress Driven Cell Death," *Toxicology Letters* 193 (2010): 209–16.

48. Scandalios, "Oxidative Stress"; McGraw et al., "Ecological Significance of Antioxidants and Oxidative Stress"; Metcalfe and Alonso-Alvarez, "Oxidative Stress as a Life-History Constraint"; Costantini, "Oxidative Stress in Ecology."

Chapter 4

1. J. Duffus, "'Heavy Metals'—A Meaningless Term?," *Pure and Applied Chemistry* 74 (2002): 793–807.

2. R. Hazen, "The Evolution of Minerals," *Scientific American* 302 (2010): 58–65.

3. Ibid.; see also chapter 3 of the present volume.

4. R. Williams, and J. Frausto da Silva, *The Chemistry of Evolution: The Development of Our Ecosystem* (New York: Elsevier, 2006), 31.

5. Ibid., 13.

6. Ibid., 29.

7. K. Kobayashi and C. Ponnamperuma, "Trace Elements in Chemical Evolution, I: Origins of Life," 16 (1985): 41–45; C. Dupont et al., "History of Biological Metal Utilization Inferred through Phylogenomic Analysis of Protein Structures," *Proceedings of the National Academy of Sciences* 107 (2010): 10567–72; H. Morowitz, V. Srinivasan, and E. Smith, "Ligand Field Theory and the Origin of Life as an Emergent Feature of the Periodic Table of Elements," *Biological Bulletin* 219 (2010): 1–6; Williams and da Silva, *Chemistry of Evolution*, 171.

8. F. Nielsen, "Evolutionary Events Culminating in Specific Minerals Becoming Essential for Life," *European Journal of Nutrition* 39 (2000): 62–66; A. Summers, "Damage Control: Regulating Defenses against Toxic Metals and Metalloids," *Current Opinion in Microbiology* 12 (2009): 138–44; N. Aras and Y. Atman, *Trace Element Analysis of Food and Diet* (London: Royal Society of Chemistry, 2006), chaps. 1 and 13; Kobayashi and Ponnamperuma, "Trace Elements in Chemical Evolution."

9. Aras and Atman, *Trace Element Analysis of Food and Diet*.

10. Ibid.; World Health Organization, Food and Agriculture Organization of the United Nations, and International Atomic Energy Agency, *Trace Elements in Human Nutrition and Health* (Geneva: World Health Organization, 1996).

11. C. Andreini, I. Bertini, and A. Rosato, "Metalloproteomes: A Bioinformatic Approach," *Accounts of Chemical Research* 42 (2009): 1471–79.

12. Kobayashi and Ponnamperuma, "Trace Elements in Chemical Evolution."

13. F. Crick and L. Orgel, "Directed Panspermia," *Icarus* 19 (1973): 344.

14. Kobayashi and Ponnamperuma, "Trace Elements in Chemical Evolution." See also references to W. Shaw, "Studies in Biogeochemistry: II. Discussion and References," *Geochimica et Cosmochimica Acta* 19 (1960): 207–15, in Nielsen, "Evolutionary Events Culminating in Specific Minerals Becoming Essential for Life"; and Williams and da Silva, *Chemistry of Evolution*.

15. J. Cavat, G. Borrelly, and N. Robinson, "Zn, Cu, and Co in Cyanobacteria: Selective Control of Metal Availability," *Federation of European Microbiological Societies* 27 (2003): 165–81.

16. For information about siderophores, see ibid.

17. Reviewed in U. Schaible and S. Kaufmann, "Iron and Microbial Infection," *Nature Reviews Microbiology* 2 (2004): 946–53.

18. A. Mulkidjanian, "On the Origin of Life in the Zinc World, 1: Photosynthesizing, Porous Edifices Build of Hydrothermally Precipitated Zinc Sulfide as Cradles of Life on Earth," *Biology Direct* 4 (2009): 26; A. Mulkidjanian

and M. Galperin, "On the Origin of Life in the Zinc World, 2: Validation of the Hypothesis on the Photosynthesizing Zinc Sulfide Edifices as Cradles of Life on Earth," *Biology Direct* 4 (2009): 27. For a different view, see C. L. Dupont and G. Caetano-Anolles, "Reply to Mulkidjanian and Galperin: Zn May Have Constrained Evolution during the Proterozoic but Not the Archean," *Proceedings of the National Academy of Sciences* 107 (2010): E138.

19. For examples, see "New Research Rejects 80-Year Theory of 'Primordial Soup' as the Origin of Life," *ScienceDaily*, February 3, 2010, accessed July 2011, http://www.sciencedaily.com/releases/2010/02/100202101245.htm; and "Deep Sea Vents—Origin of Life?," *Pulse of the Planet*, August 19, 1998, accessed June 2011, http://www.pulseplanet.com/dailyprogram/dailies.php ?POP=1688.

20. Mulkidjanian, "On the Origin of Life in the Zinc Worlds"; Mulkidjanian and Galperin, "On the Origin of Life in the Zinc Worlds: 2."

21. Dupont and Caetano-Anolles, "Reply to Mulkidjanian and Galperin."

22. C. Dupont et al., "Modern Proteomes Contain Putative Imprints of Ancient Shifts in Trace Metal Geochemistry," *Proceedings of the National Academy of Sciences* 103 (2006): 17822–27.

23. Ibid.; Dupont et al., "History of Biological Metal Utilization."

24. Summers, "Damage Control"; R. Goyer, "Toxic and Essential Metal Interactions," *Annual Review of Nutrition* 17 (1997): 37–50.

25. M. Cuillel, "The Dual Personality of Ionic Copper in Biology," *Journal of Inclusion Phenomena and Macrocyclic Chemistry* 65 (2009): 165–70.

26. Dupont et al., "Modern Proteomes Contain Putative Imprints of Ancient Shifts in Trace Metal Geochemistry"; Dupont et al., "History of Biological Metal Utilization."

27. Cuillel, "Dual Personality of Ionic Copper in Biology."

28. A. Rosenzweig, "Metallochaperones: Bind and Deliver," *Chemistry and Biology* 9 (2002): 673–77; Andreini, Bertini, and Rosato, "Metalloproteomes"; Dupont et al., "History of Biological Metal Utilization."

29. H.-F. Ji et al., "Evolutionary Formation of New Protein Folds Is Linked to Metallic Cofactor Recruitment," *BioEssays* 31 (2009): 975–80.

30. For one of the earliest descriptions of MT, see M. Margoshes and B. L. Vallee, "A Cadmium Protein from Equine Kidney Cortex," *Journal of the American Chemical Society* 79 (1957): 4813–14.

31. Reviewed in M. Nordberg and G. Nordberg, "Metallothioneins: Historical Development and Overview," *Metal Ions in Life Sciences* 5 (2009): 1–29; P. Coyle et al., "Metallothionein: The Multipurpose Protein," *Cellular and Molecular Life Sciences* 59 (2002): 627–47.

32. Nordberg and Nordberg, "Metallothioneins."

33. Coyle et al., "Metallothionein."

34. Reviewed in J. Hidalgo et al., "Structure and Function of Vertebrate Metallothioneins," *Metal Ions in Life Sciences* 5 (2009): 279–317. D.-H. Nam

et al., "Molecular Characterization of Two Metallothionein Isoforms in Avian Species: Evolutionary History, Tissue Distribution Profile, and Expression Associated with Metal Accumulation," *Comparative Biochemistry and Physiology* 145 (2007): 295–305.

35. Nam et al., "Molecular Characterization of Two Metallothionein Isoforms in Avian Species."

36. Ibid.

37. P.-A. Binz and J. H. R. Kägi, "Molecular Evolution of Metallothioneins: Contributions from Coding and Non-coding Regions," poster presented at the Second European Meeting of the Protein Society in Cambridge (UK), 1997, accessed June 2011, http://www.bioc.unizh.ch/mtpage/poster/posterevol.html.

38. T. Janssens, D. Roelofs, and N. van Straalen, "Molecular Mechanisms of Heavy Metal Tolerance and Evolution in Insects," *Insect Science* 16 (2009): 3–18; A. Miles et al., "Induction, Regulation, Degradation, and Biological Significance of Mammalian Metallothioneins," *Critical Reviews in Biochemistry and Molecular Biology* 35 (2000): 35–70.

39. F. Trinchella et al., "Molecular Cloning and Sequencing of Metallothionein in Squamates: New Insights into the Evolution of the Metallothionein Genes in Vertebrates," *Gene* 423 (2008): 48–56.

40. Janssens, Roelofs, and Straalen, "Molecular Mechanisms of Heavy Metal Tolerance and Evolution in Insects."

41. J. Gutierrez, F. Amaro, and A. Martin-Gonzalez, "From Heavy Metal-Binders to Biosensors: Ciliate Metallothioneins Discussed," *BioEssays* 31 (2009): 805–16; Trinchella et al., "Molecular Cloning and Sequencing of Metallothionein in Squamates."

42. M. Soskine and D. Towfik, "Mutational Effects and the Evolution of New Protein Functions," *Nature Review Genetics* 11 (2010): 572–82.

43. N. van Straalen and D. Roelofs, *Introduction to Ecological Genomics* (Oxford: Oxford University Press, 2006), 62.

44. Nam et al., "Molecular Characterization of Two Metallothionein Isoforms in Avian Species"; Trinchella et al., "Molecular Cloning and Sequencing of Metallothionein in Squamates"; Hidalgo et al., "Structure and Function of Vertebrate Metallothioneins."

45. Janssens, Roelofs, and Straalen, "Molecular Mechanisms of Heavy Metal Tolerance and Evolution in Insects."

46. Ibid.

47. Gutierrez, Amaro, and Martin-Gonzalez, "From Heavy Metal-Binders to Biosensors"; M. Valls et al., "A New Insight in Metallothionein (MT) Classification and Evolution," *Journal of Biological Chemistry* 276 (2001): 32835–43.

48. Reviewed in C. Blindauer and O. Leszczyszyn, "Metallothioneins: Unparalleled Diversity in Structures and Functions for Metal Ion Homeostasis and More," *Natural Product Reports* 27 (2009): 720–41.

49. C. Ouzounis et al., "A Minimal Estimate for the Gene Content of the Last Universal Common Ancestor—Exobiology from a Terrestrial Perspective," *Research in Microbiology* 157 (2006): 57–68.

50. Cavat, Borrelly, and Robinson, "Zn, Cu, and Co in Cyanobacteria."

51. Hidalgo et al., "Structure and Function of Vertebrate Metallothioneins"; Coyle et al., "Metallothionein."

52. C. Klaassen, J. Liu, and S. Choudhuri, "Metallothionein: An Intracellular Protein to Protect against Cadmium Toxicity," *Annual Review of Pharmacology and Toxicology* 39 (1999): 267–94.

53. European Food Safety Authority, "Cadmium in Food: Scientific Opinion of the Panel on Contaminants in the Food Chain," *European Food Safety Journal* 980 (2009): 1–139.

54. Nikhil Johri, Gregory Jacquillet, and Robert Unwin, "Heavy Metal Poisoning: The Effects of Cadmium on the Kidney," *Biometals* 23 (2010): 783–92.

55. For the remainder of this discussion it will be referred to as metallothionein, as suggested following personal communication with Paul Klerks.

56. P. Klerks and J. Levinton, "Rapid Evolution of Metal Resistance in a Benthic Oligochaete Inhabiting a Metal-Polluted Site," *Biological Bulletin* 176 (1989): 135–41.

57. J. Levinton et al., "Rapid Loss of Genetically Based Resistance to Metals after the Cleanup of a Superfund Site," *Proceedings of the National Academy of Sciences of the United States of America* 100 (2003): 9889–91.

58. One caveat is that laboratory-induced resistance could have been different than field-induced resistance.

59. Joshua Mackie et al., "Loss of Evolutionary Resistance by the Oligochaete *Limnodrilus hoffmeisteri* to a Toxic Substance—Cost or Gene Flow?," *Evolution* 64 (2010): 152–65.

60. Reviewed in R. D. Barrett and D. Schluter, "Adaptation from Standing Genetic Variation," *Trends in Ecology and Evolution* 23 (2008): 38–44.

61. Coyle et al., "Metallothionein," 640.

Chapter 5

1. For many years, it was thought that some eukaryotes, including the protist *Giardia*, were lacking mitochondria, but subsequent studies now indicate that "all eukaryotes appear to contain either a mitochondrion, a hydrogenosome, a mitosome, or a mitochondrion-like organelle (MLO), all of which resulted from a single, ancient, endosymbiotic event." A. Shiflett and P. Johnson, "Mitochondrion-Related Organelles in Eukaryotic Protists," *Annual Review in Microbiology* 64 (2010): 423.

2. For a review of eukaryote evolutionary history, see T. Embley and W. Martin, "Eukaryotic Evolution, Changes, and Challenges," *Nature* 440 (2006): 623–30.

3. N. Lane, *Oxygen: The Molecule That Made the World* (Oxford: Oxford University Press, 2009), 51.

4. One exception is the recent discovery of deep-sea-dwelling animals belonging to the phylum *Loricifera*, which thrive under anoxic conditions. For more, see R. Danovaro et al., "The First Metazoa Living in Permanently Anoxic Conditions," *BioMed Central Biology* 8 (2010): 30.

5. For a review, see R. Grosberg and R. Strathmann, "The Evolution of Multicellularity: A Minor Major Transition?," *Annual Review of Ecology, Evolution, and Systematics* 38 (2007): 621–54.

6. A. Brown and A. Galea, "Cholesterol as an Evolutionary Response to Living with Oxygen," *Evolution: International Journal of Organic Evolution* 64 (2010): 2179–83; J. Saul and L. Schwartz, "Cancer as a Consequence of the Rising Level of Oxygen in the Late Precambrian," *Lethaia* 40 (2007): 211–20.

7. M. Lynch, "The Origins of Eukaryotic Gene Structure," *Molecular Biology and Evolution* 23 (2006): 450–68.

8. Ibid., 464.

9. N. Lane and W. Martin, "The Energetics of Genome Complexity," *Nature* 467 (2010): 929.

10. There are at least two dozen cases where prokaryotic or eukaryotic unicellular life independently initiated a path toward a more social life, and some were more successfully than others. Examples of multicellular prokaryotes include filamentous cyanobacteria observed in fossils dating back over three billion years, and the odd myxobacteria—which travel in swarms, forming odd-shaped fruiting bodies during reproduction—and the recently identified magnetotactic multicellular prokaryotes, or MMP, which not only reproduce as a whole, but also have yet to be found in unicellular form.

11. For many years, sponges were excluded from categorization as animals because they lack specialized digestive tracts and neurological development.

12. M. Srivastava et al., "The *Amphimedon queenslandica* Genome and the Evolution of Animal Complexity," *Nature* 466 (2010): 720–26; A. Abderrazak et al., "Large Colonial Organisms with Coordinated Growth in Oxygenated Environments 2.1_Gyr Ago," *Nature* 466 (2010): 100–104.

13. A. Mann, "Sponge Genome Goes Deep," *Nature News* 466 (2010): 673; A. Rokas, "The Origins of Multicellularity and the Early History of the Genetic Toolkit for Animal Development," *Annual Review of Genetics* 42 (2008): 235–51; Srivastava et al., "The *Amphimedon queenslandica* Genome and the Evolution of Animal Complexity."

14. For an interesting discussion of cell suicide and cancer defense, see Mann, "Sponge Genome Goes Deep."

15. T. Domazet-Loso and D. Tautz, "Phylostratigraphic Tracking of Cancer Genes Suggests a Link to the Emergence of Multicellularity in Metazoa," *BioMed Central Biology* 8 (2010): 66; F. Ciccarelli, "The (R)evolution of Cancer Genetics," *BioMed Central Biology* 8 (2010): 74.

16. J. Robert, "Comparative Study of Tumorigenesis and Tumor Immunity in Invertebrates and Nonmammalian Vertebrates," *Developmental and Comparative Immunology* 34 (2010): 915–25.

17. For a detailed review of human cancer history, see M. Greaves, *Cancer: The Evolutionary Legacy* (Oxford: Oxford University Press, 2000); and S. Mukherjee, *Emperor of All Maladies* (New York: Scribner, 2010).

18. D. Wheatley, "Carcinogenesis: Is There a General Theorem?," *BioEssays* 5 (2010): 111; C. Sonnenschein and A. Soto, "Theories of Carcinogenesis: An Emerging Perspective," *Seminars in Cancer Biology* 18 (2009): 372–77.

19. Greaves, *Cancer*, 23.

20. For discussions of alternative theories of cancer, see Wheatley, "Carcinogenesis"; and Sonnenschein and Soto, "Theories of Carcinogenesis."

21. D. Hanahan and R. Weinberg, "The Hallmarks of Cancer," *Cell* 100 (2000): 57.

22. M. Stratton, P. Campbell, and P. Futreal, "The Cancer Genome," *Nature* 458 (2009): 719–24.

23. B. Crespi and K. Summers, "Evolutionary Biology of Cancer," *Trends in Ecology and Evolution* 20 (2005): 545–52.

24. Ibid., 547. Note that this quote refers to a citation by N. Komarova and D. Wodarz, "Evolutionary Dynamics of Mutator Phenotypes in Cancer: Implications for Chemotherapy," *Cancer Research* 63 (2003): 6635–42.

25. For a review and analysis of genes associated with cancer, including caretaker and gatekeeper genes, see Domazet-Loso and Tautz, "Phylostratigraphic Tracking of Cancer Genes"; and N. Levitt and I. Hickson, "Caretaker Tumour Suppressor Genes That Defend Genome Integrity," *Trends in Molecular Medicine* 8 (2002): 179–86.

26. M. Junttila and G. Evan, "p53—a Jack of All Trades but Master of None," *Nature Reviews Cancer* 9 (2009): 821–29.

27. For a recent discussion of the importance of p63 and p73, see P. Kiberstis and E. Marshall, "Celebrating an Anniversary," *Science* 331 (2011): 1539.

28. N. King et al., "The Genome of the Choanoflagellate *Monosiga brevicollis* and the Origin of Metazoans," *Nature* 451 (2008): 783–88; R. Rutkowski, K. Hofmann, and A. Gartner, "Phylogeny and Function of the Invertebrate p53 Superfamily," *Cold Spring Harbor: Perspectives in Biology* 2 (2010): a001131.

29. Rutkowski, Hofmann, and Gartner, "Phylogeny and Function of the Invertebrate p53 Superfamily"; V. Belyi and A. Levine, "One Billion Years of p53/p63/p73 Evolution," *Proceedings of the National Academy of Sciences of the United States of America* 106 (2009): 17609–10; V. Belyi et al., "The Origins and Evolution of the p53 Family of Genes," *Cold Spring Harbor: Perspectives in Biology* 2 (2010): a001198.

30. Belyi et al., "The Origins and Evolution of the p53 Family of Genes."

31. Srivastava et al., "The *Amphimedon queenslandica* Genome and the Evolution of Animal Complexity."

32. "In Journey from Maggot to Fruit Fly, a Clue about Cancer Metastasis," *ScienceDaily*, January 21, 2010, accessed June 2011, http://www.science daily.com/releases/2010/01/100119121208.htm. For a discussion of specific genes and cancer in fruit flies, see T. Donaldson and R. Duronio, "Cancer Cell Biology: Myc Wins the Competition," *Current Biology* 14 (2004): R425–R27.

33. See W.-J. Lu, J. Amatruda, and J. Abrams, "p53 Ancestry: Gazing through an Evolutionary Lens," *Nature Reviews Cancer* 9 (2009): 758–62; and Belyi at al., "The Origins and Evolution of the p53 Family of Genes" (for phylogeny of the p53 gene family). The research on fruit fly brain cells was done by Johannes Baur; for more, see "Key to Longer Life (in Fruit Flies) Lies in Just 14 Brain Cells," *Brown University News and Events*, September 20, 2007, accessed July 2011, http://news.brown.edu/pressreleases/2007/09/key-longer-life.

34. Reviewed in Belyi et al., "The Origins and Evolution of the p53 Family of Genes"; see also A. Muttray, P. Schulte, and S. Baldwin, "Invertebrate p53—Like mRNA Isoforms Are Differentially Expressed in Mussel Haemic Neoplasia," *Marine Environmental Research* 66 (2008): 412–21.

35. Belyi et al., "The Origins and Evolution of the p53 Family of Genes"; Rutkowski, Hofmann, and Gartner, "Phylogeny and Function of the Invertebrate p53 Superfamily."

36. N. Nakamuta and S. Kobayashi, "Expression of p63 in the Mouse Ovary," *Journal of Reproduction and Development* 53 (2007): 691–97; Rutkowski, Hofmann, and Gartner, "Phylogeny and Function of the Invertebrate p53 Superfamily"; G. Blandino and M. Dobbelstein, "p73 and p63: Why Do We Still Need Them?," *Cell Cycle* 3 (2004): 886–94. For more about their possible role in cancer cells, see Kiberstis and Marshall, "Celebrating an Anniversary."

37. Junttila and Evan, "p53—a Jack of All Trades but Master of None."

38. For a discussion of the relationship between zinc and cancer, see S. Loh, "The Missing Zinc: p53 Misfolding and Cancer," *Metallomics* 2 (2010): 442–49. According to Loh, regarding the importance of zinc and p53, "(a) failure of p53 to fold leads to loss of function, (b) loss of p53 function is one of the things that is strongly linked to cancer (it happens in half of all tumors), (c) many common cancer-promoting mutations in p53 act by causing the protein to misfold. In vitro, p53 folds best in the presence of just the right amount of zinc. By that I mean exactly one zinc ion per protein molecule, or (more physiologically relevant) in the presence of a metallochaperone and zinc. p53 can fold quite well in the absence of zinc, but the zinc-free form of the protein isn't very stable and it would probably aggregate or be cleared by housekeeping enzymes in the cell. The metallochaperone is there to

hand p53 a zinc ion as soon as it's ready to accept one (i.e., when the native binding site has formed). p53 has a much harder hard time folding when too much zinc is around. Zinc ends up binding weakly to the wrong ligands, of which there are many, and holding the protein in the wrong conformation so that it can't fold properly. The metallochaperone is also there to rein in the zinc when the protein is trying to fold and is vulnerable to misligation. This is presumably one of the reasons why free metal ions are kept at very low concentration by metallothioneins and other metal chaperones in the cell (another reason being their ability to generate harmful reactive oxygen species)." Stuart Loh, e-mail exchange with author, December 10, 2010.

39. For a discussion of metallothionein and cancer, see T. Eckschlager et al., "Metallothioneins and Cancer," *Current Protein and Peptide Science* 10 (2009): 360–75.

40. Junttila and Evan, "p53—a Jack of All Trades but Master of None," 823.

41. Ibid., 826.

Chapter 6

1. This chapter could have also focused on toxic chemicals used by animals, and there are many good sources for those interested in this topic. See, for example, B. Fry et al., "The Toxicogenomic Multiverse: Convergent Recruitment of Proteins into Animal Venoms," *Annual Review of Genomics and Human Genetics* 10 (2009): 483–511.

2. Discussed in R. Feyereisen, "Arthropod CYPomes Illustrate the Tempo and Mode in P450 Evolution," *Biochimica et Biophysica Acta* 1814 (2011): 19–28.

3. T. Lefèvre et al., "Evidence for Trans-generational Medication in Nature," *Ecology Letters* 13 (2010): 1485–93.

4. "All Land Plants Evolved from Single Type of Algae, Scientists Say," *Charlestown Daily Mail*, June 4, 2001, accessed August 2011, http://news.nationalgeographic.com/news/2001/06/0604_wirealgae.html; R. Chapman and D. Waters, "Green Algae and Land Plants—an Answer at Last?," *Journal of Phycology* 240 (2002): 237–40; P. Kenrick, "Fishing for the First Plants," *Nature* 425 (2003): 248–49.

5. For a review, see P. Gensel, "The Earliest Land Plants," *Annual Review of Ecology, Evolution, and Systematics* 39 (2008): 459–77.

6. E. Pacini, L. Viegi, and G. G. Franchi, "Types, Evolution and Significance of Plant-Animal Interactions," *Rendiconti Lincei* 19 (2008): 75–101.

7. Ina Schaefer et al., "Arthropod Colonization of Land-Linking Molecules and Fossils in Oribatid Mites (*Acari, Oribatida*)," *Molecular Phylogenetics and Evolution* 57 (2010): 113–21.

8. Ibid.

9. For those interested in more detail on evolution in general, and the colonization of land in particular, see R. Dawkins, *The Ancestor's Tale: A Pilgrimage to the Dawn of Evolution* (New York: Houghton Mifflin, 2004).

10. Sea squirts or tunicates are defined in part by their rodlike notochord and hollow nerve tube, during at least one stage of their life.

11. For a review, see H. D. Sues and R. R. Reisz, "Origins and Early Evolution of Herbivory in Tetrapods," *Trends in Ecology and Evolution* 13 (1988): 141–45.

12. Ibid.

13. R. Thomas, N. Sah, and P. Sharma, "Therapeutic Biology of *Jatropha curcas*: A Mini Review," *Current Pharmaceutical Biotechnology* 9 (2008): 315–24.

14. C. Theoduloz et al., "Antiproliferative Activity of the Diterpenes Jatrophone and Jatropholone and Their Derivatives," *Planta Medica* 75 (2009): 1520–22.

15. J. Landsberg et al., "Saxitoxin Puffer Fish Poisoning in the United States, with the First Report of *Pyrodinium bahamense* as the Putative Toxin Source," *Environmental Health Perspectives* 114 (2006): 1502–7.

16. For more about the origins of saxitoxin, see S. Murray, T. K. Mihali, and B. Neilan, "Extraordinary Conservation, Gene Loss, and Positive Selection in the Evolution of an Ancient Neurotoxin," *Molecular Biology* 28 (2011): 1173–82.

17. *Chemical Information Review Document for L-beta-Methylaminoalanine* (Research Triangle Park, NC: National Toxicology Program, 2008), accessed June 2011, http://ntp.niehs.nih.gov/ntp/htdocs/Chem_Background/ExSum Pdf/LbetaMethylaminoalanine_508.pdf.

18. For a review, see L. Snyder and T. E Marler, "Rethinking Cycad Metabolite Research," *Communicative and Integrative Biology* 4 (2011): 86–88.

19. J. Duke, "Medicinal Plants and the Pharmaceutical Industry," in *New Crops*, ed. J. Janick and J. E. Simon (New York: Wiley, 1993), 664–69.

20. M. Wink, "Evolution of Secondary Metabolites from an Ecological and Molecular Phylogenetic Perspective," *Phytochemistry* 64 (2003): 3–19.

21. Ibid.; M. Wink, "Plant Secondary Metabolism: Diversity, Function and Its Evolution," *Natural Products Communications* 8 (2008): 1205–16.

22. Wink, "Evolution of Secondary Metabolites," 5.

23. D. Samuel, "A Review of the Effects of Plant Estrogenic Substances on Animal Reproduction," *Reproduction* 67 (1967): 308–12.

24. Wink, "Evolution of Secondary Metabolites," 5.

25. For a review of the paradox of drug rewards, see R. Sullivan, E. Hagen, and P. Hammerstein, "Revealing the Paradox of Drug Reward in Human Evolution," *Proceedings of the Royal Society of London* 275 (2008): 1231–41.

26. T. W. Soong and B. Venkatesh, "Adaptive Evolution of Tetrodotoxin Resistance in Animals," *Trends in Genetics* 22 (2006): 621–26.

27. J. Johnston, "Evaluation of Cocoa- and Coffee-Derived Methylxanthines as Toxicants for the Control of Pest Coyotes," *Journal of Agricultural and Food Chemistry* 53 (2005): 4069–75.

28. For more about Mithridates, see A. Mayer, *The Poison King: The Life and Legend of Mithridates, Rome's Deadliest Enemy* (Princeton, NJ: Princeton University Press, 2009).

29. J. Stegeman, personal communication, February 16, 2011.

30. D. Nelson, "The Cytochrome P450 Home Page," *Human Genomics* 4 (2009): 59–65; also available online, accessed June 2011, http://drnelson .uthsc.edu/P450.statistics.Aug2009.pdf.

31. D. W. Nebert and M. Z. Dieter, "The Evolution of Drug Metabolism," *Pharmacology* 61 (2000): 124–35.

32. R. Estabrook, "A Passion for P450s (Remembrances of the Early History of Research on Cytochrome P450)," *Drug Metabolism and Disposition* 31 (2003): 1461–73.

33. D. Werck-Reichhart and R. Feyereisen, "Cytochromes P450: A Success Story," *Genome Biology* 1 (2000): 1–9.

34. Y. Yoshida et al., "Sterol 14-Demethylase P450 (CYP51) Provides a Breakthrough for the Discussion on the Evolution of Cytochrome P450 Gene Superfamily," *Biochemical and Biophysical Research Communications* 273 (2000): 799–804; and X. Qi et al., "A Different Function for a Member of an Ancient and Highly Conserved Cytochrome P450 Family: From Essential Sterols to Plant Defense," *Proceedings of the National Academy of Sciences of the United States of America* 103 (2006): 18848–53; G. Lepesheva and M. Waterman, "CYP51-the Omnipotent P450," *Molecular and Cellular Endocrinology* 215 (2004): 165–70.

35. C. Jackson et al., "P450s in Microbial Sterol Biosynthesis and Drug Targets and the Formation of Secondary Metabolites," *Acta Chimica Slovenica* 55 (2008): 58–62.

36. Mark Hahn, personal communication, June 16, 2011.

37. D. Nelson, "Progress in Tracing the Evolutionary Paths of Cytochrome P450," *Biochimica et Biophysica Acta* 1814 (2011): 14–18.

38. Feyereisen, "Arthropod CYPomes Illustrate the Tempo and Mode in P450 Evolution," 19.

39. For those interested in CYP naming, see D. Nelson, "Notes on Nomenclature," n.d., accessed September 2011, http://drnelson.uthsc.edu /naming.html.

40. R. Feyereisen, "Evolution of Insect P450," *Biochemical Society Transactions* 34 (2006): 1252–55.

41. J. Goldstone et al., "Cytochrome P450 1 Genes in Early Deuterostomes (Tunicates and Sea Urchins) and Vertebrates (Chicken and Frog): Origin and Diversification of the CYP1 Gene Family," *Molecular Biology and Evolution* 24 (2007): 2619–31.

42. For a review of CYPs role in cancer, see D. Nebert and T. Dalton, "The Role of Cytochrome P450 Enzymes in Endogenous Signalling Pathways and Environmental Carcinogenesis," *Nature Reviews Cancer* 6 (2006): 947–60.

43. F. J. Gonzalez and D. W. Nebert, "Evolution of the P450 Gene Superfamily: Animal-Plant 'Warfare,' Molecular Drive, and Human Genetic Differences in Drug Oxidation," *Trends in Genetics* 6 (1990): 182–86.

44. J. Goldstone, e-mail communication, January 7, 2011.

45. There is evidence that umbellifers in forests do not produce photoactivated chemicals.

46. M. Berenbaum and P. Feeny, "Toxicity of Angular Furanocoumarins to Swallowtail Butterflies: Escalation in a Coevolutionary Arms Race?," *Science* 212 (1981): 927.

47. Reviewed in M. Schuler, "P450s in Plant-Insect Interactions," *Biochimica et Biophysica Acta* 1814 (2011): 36–45; M. Berenbaum, C. Favret, and M. Schuler, "On Defining 'Key Innovations' in an Adaptive Radiation: Cytochrome P450S and Papilionidae," *American Naturalist* 148 (2011): S139–S155.

48. M. Berenbaum, "Coumarins and Caterpillars: A Case for Coevolution," *Evolution* 37 (1983): 173.

49. Ibid.

50. Zhimou Wen et al., "CYP6B1 and CYP6B3 of the Black Swallowtail (*Papilio polyxenes*): Adaptive Evolution through Subfunctionalization," *Molecular Biology and Evolution* 23 (2006): 2434–43.

51. See Schuler, "P450s in Plant-Insect Interactions"; and Berenbaum, Favret, and Schuler, "On Defining 'Key Innovations' in an Adaptive Radiation."

52. Feyereisen, "Arthropod CYPomes Illustrate the Tempo and Mode in P450 Evolution."

53. Wen et al., "CYP6B1 and CYP6B3 of the Black Swallowtail."

54. J. Goldstone and J. Stegeman, personal communication, February 16, 2011.

55. Reviewed in Wen et al., "CYP6B1 and CYP6B3 of the Black Swallowtail."

56. J. G. Scott and Z. Wen, "Cytochromes P450 of Insects: The Tip of the Iceberg," *Pest Management Science* 57 (2001): 958–67.

57. For a review of plant resistance, see J. Gunsolus, "Herbicide Resistant Weeds," Regents of the University of Minnesota, 2002, accessed August 2011, http://www.extension.umn.edu/distribution/cropsystems/dc6077.html. For a review of insecticide resistance in insects, see J. Forgash, "History, Evolution and Consequences of Insecticide Resistance," *Pesticide Biochemistry and Biophysiology* 22 (1984): 178–86; see also W. Allen, *The War on Bugs* (White River Junction, VT: Chelsea Green, 2008).

58. See P. Tandale, C. Meena, and R. Gaud, "Drug Interactions and Grapefruit Juice," Pharmainfo.net, February 2, 2007, accessed June 2011, http://www.pharmainfo.net/reviews/drug-interactions-grapefruit-juice; E. Monosson, "More on Grapefruit Juice and Drugs," *Neighborhood Toxicologist*, November 26, 2007, accessed June 2011, http://theneighborhood toxicologist.blogspot.com/2007/11/more-on-grapefruit-juice-and-drugs .html.

59. Tandale, Meena, and Gaud, "Drug Interactions and Grapefruit Juice"; Monosson, "More on Grapefruit Juice and Drugs."

60. S. P. Westphal, "Grapefruit Effect on Drug Levels Has Sweeter Side," *Wall Street Journal*, November 27, 2007, accessed June 2011, http://online .wsj.com/article/SB119612613623804707.html.

61. M. Berenbaum, "Postgenomic Chemical Ecology: From Genetic Code to Ecological Interactions," *Journal of Chemical Ecology* 28 (2002): 878.

Chapter 7

1. L. Emberson et al., "Overheard Cell-Phone Conversations: When Less Speech Is More Distracting," *Psychological Sciences* 21 (2010): 1383–88.

2. R. Nicoll and B. Alger, "The Brain's Own Marijuana," *Scientific American* 291 (2004): 68–75.

3. G. Burnstock and A. Verkhratsky, "Evolutionary Origins of the Purinergic Signalling System," *Acta Physiologica* 195 (2009): 415–47.

4. B. Khakh and G. Burnstock, "The Double Life of ATP," *Scientific American* 6 (2009): 84–92.

5. Burnstock and Verkhratsky, "Evolutionary Origins of the Purinergic Signalling System," 435.

6. Khakh and Burnstock, "Double Life of ATP."

7. G. Markov and V. Laudet, "Origin and Evolution of the Ligand-Binding Ability of Nuclear Receptors," *Molecular and Cellular Endocrinology* 334 (2011): 21–30.

8. M. Robinson-Rechavi et al., "How Many Nuclear Hormone Receptors Are There in the Human Genome?," *Trends in Genetics* 17 (2001): 554–56.

9. J. Bridgham et al., "Protein Evolution by Molecular Tinkering: Diversification of the Nuclear Receptor Superfamily from a Ligand-Dependent Ancestor," *PLoS Biology* 8 (2010): e1000497; A. Reitzel et al., "Nuclear Receptors from the Ctenophore *Mnemiopsis leidyi* Lack a Zinc-Finger DNA-Binding Domain: Lineage-Specific Loss or Ancestral Condition in the Emergence of the Nuclear Receptor Superfamily?," *EvoDevo* 2 (2011): 3; J. W. Thornton, "Nonmammalian Nuclear Receptors: Evolution and Endocrine Disruption," *Pure and Applied Chemistry* 75 (2003): 1827–39.

10. Bridgham et al., "Protein Evolution by Molecular Tinkering"; Reitzel et al., "Nuclear Receptors from the Ctenophore *Mnemiopsis leidyi*"; Thornton, "Nonmammalian Nuclear Receptors: Evolution and Endocrine Disruption"; H. Escriva, F. Delaunay, and V. Laudet, "Ligand Binding and Nuclear Receptor Evolution," *BioEssays* 22 (2000): 717–27; P. Dehal and J. Boore, "Two Rounds of Whole Genome Duplication in the Ancestral Vertebrate," *PLoS Biology* 3 (2005): e314. For a synopsis of the above article, see Liza Gross, "Clear Evidence for Two Rounds of Vertebrate Genome Duplication," *PLoS Biology* 3 (2005): e344.

11. Escriva, Delaunay, and Laudet, "Ligand Binding and Nuclear Receptor Evolution"; Bridgham et al., "Protein Evolution by Molecular Tinkering"; Reitzel et al., "Nuclear Receptors from the Ctenophore."

12. Escriva et al., "Ligand Binding and Nuclear Receptor Evolution."

13. Reviewed in Bridgham et al., "Protein Evolution by Molecular Tinkering."

14. Markov and Laudet, "Origin and Evolution of the Ligand-Binding Ability of Nuclear Receptors."

15. Reitzel et al., "Nuclear Receptors from the Ctenophore"; Thornton, "Nonmammalian Nuclear Receptors." For a review of ligand evolution, see Frances M. Sladek, "What Are Nuclear Receptor Ligands?," *Molecular and Cellular Endocrinology* 334 (2010): 1–11.

16. Reitzel et al., "Nuclear Receptors from the Ctenophore." Reitzel proposes a model evolutionary diversification of the NR superfamily in animals before the split between bilaterans and radially symmetrical animals.

17. For more on Theo Colborn's work, see "The Endocrine Disruption Exchange," n.d., accessed June 2011, http://www.endocrinedisruption.com/; and T. Colborn, D. Dumanowski, and P. Meyers, *Our Stolen Future* (New York: Penguin, 1996).

18. L. Gray et al., "Environmental Antiandrogens: Low Doses of the Fungicide Vinclozolin Alter Sexual Differentiation of the Male Rat," *Toxicology and Industrial Health* 15 (1999): 48–64.

19. For more on androgen insensitivity, see "Androgen Insensitivity Syndrome," *Genetics Home Reference*, May 2008, accessed June 2011, http://ghr.nlm.nih.gov/condition/androgen-insensitivity-syndrome.

20. Gray et al., "Environmental Antiandrogens." More recently, water contaminated with environmentally relevant amounts of vinclozolin was found to interfere with normal male behavior in frogs. For more, see F. Hoffmann and W. Kloas, "An Environmentally Relevant Endocrine-Disrupting Antiandrogen, Vinclozolin, Affects Calling Behavior of Male *Xenopus laevis*," *Hormones and Behavior* 58 (2010): 653–59.

21. W. Kelce et al., "Persistent DDT Metabolite p,p'–DDE Is a Potent Androgen Receptor Antagonist," *Nature* 375 (1995): 581–85.

22. For more on endocrine disruption, see "The Endocrine Disruption Exchange," n.d., accessed August 2011, http://www.endocrinedisruption .com/; and Colborn, Dumanowski, and Meyers, *Our Stolen Future*. For one of the first compilations of research across species, see T. Colborn and C. Clement, eds., *Chemically-Induced Alterations in Sexual and Functional Development: The Wildlife/Human Connection* (Princeton, NJ: Princeton Scientific Publishing, 1992).

23. Reviewed in G. Eick and J. Thornton, "Evolution of Steroid Receptors from an Estrogen-Sensitive Ancestral Receptor," *Molecular and Cellular Endocrinology* 334 (2011): 31–38. For a chart, see "Nuclear Receptor," *Wikipedia*, n.d., accessed August 2011, http://en.wikipedia.org/wiki/Nuclear _receptor.

24. Eick and Thornton, "Evolution of Steroid Receptors," 32.

25. Ibid.

26. Ibid.

27. Markov and Laudet, "Origin and Evolution of the Ligand-Binding Ability of Nuclear Receptors."

28. Thornton, "Nonmammalian Nuclear Receptors," 1832.

29. Ibid.

30. Ibid. (emphasis added).

31. A good source of information on endocrine disruption chemicals and receptors can be found at "e.hormone," *EDC Sources*, 2011, accessed June 2011, http://e.hormone.tulane.edu/learning/sources.html.

32. The plastic bisphenol A, for example, was once under consideration for use *as* a synthetic estrogen. For more about BPA, see E. Monosson, "A Great Future in Plastics," *Neighborhood Toxicologist*, May 20, 2008, accessed June 2011, http://theneighborhoodtoxicologist.blogspot.com/2008 /05/great-future-in-plastics.html; and E. Monosson, "Polycarbonate Redux," *Neighborhood Toxicologist*, April 16, 2008, accessed June 2011, http://the neighborhoodtoxicologist.blogspot.com/2008/04/polycarbonate-redux .html.

33. L. Earl Gray, personal communication, February 7, 2011.

34. J. Goldstone, personal communication, February 16, 2011.

35. Q. Ma, "Xenobiotic-Activated Receptors: From Transcription to Drug Metabolism to Disease," *Chemical Research in Toxicology* 21 (2008): 1651–71.

36. For a review, see J. Raloff, "Redefining Dioxins," *Science News* 155 (1999): 156–57.

37. A. Okey, "An Aryl Hydrocarbon Receptor Odyssey to the Shores of Toxicology: The Deichmann Lecture, International Congress of Toxicology-XI," *Toxicological Sciences* 98 (2007): 25.

38. For reviews, see M. Hahn, "Aryl Hydrocarbon Receptors: Diversity

and Evolution," *Chemico-Biological Interactions* 141 (2002): 131–60; and M. Hahn, A. Lenka, and D. Sherr, "Regulation of Constitutive and Inducible AHR Signaling: Complex Interactions Involving the AHR Repressor," *Biochemical Pharmacology* 77 (2009): 485–97.

39. Reviewed in Hahn, "Aryl Hydrocarbon Receptors."

40. M. Hahn and S. Karchner, "Structural and Functional Diversification of AHRs during Metazoan Evolution," in *The AH Receptor in Biology and Toxicology*, ed. R. Pohjanvirta (New York: Wiley, forthcoming).

41. For a review, see Hahn, "Aryl Hydrocarbon Receptors"; and M. Hahn, Lenka, and Sherr, "Regulation of Constitutive and Inducible AHR Signaling."

42. Hahn, "Aryl Hydrocarbon Receptors," 149.

43. Several endogenous ligands have recently been identified, which may help reveal roles for the AhR other than detoxification. For more, see G. Chowdhury et al., "Structural Identification of Diindole Agonists of the Aryl Hydrocarbon Receptor Derived from Degradation of Indole-3-Pyruvic Acid," *Chemical Research in Toxicology* 22 (2009): 1905–12.

44. C. Flaveny et al., "Ligand Selectivity and Gene Regulation by the Human Aryl Hydrocarbon Receptor in Transgenic Mice," *Pharmacology* 75 (2009): 1412–20; Hahn and Karchner, "Structural and Functional Diversification of AHRs."

45. Flaveny et al., "Ligand Selectivity and Gene Regulation"; Hahn and Karchner, "Structural and Functional Diversification of AHRs."

46. Flaveny et al., "Ligand Selectivity and Gene Regulation."

47. For more about xenobiotic-activated receptors, see Ma, "Xenobiotic-Activated Receptors."

Chapter 8

1. Further, there is very little correlation between genome size and number of protein-coding genes. For example, humans have roughly three billion DNA base pairs, while nematode worms have only ninety-seven million base pairs, despite the relatively small differences in gene numbers. See L. Pray, "Eukaryotic Genome Complexity," *Nature Education* 1 (2008): 1.

2. A. Gasch, "The Environmental Stress Response: A Common Yeast Response to Diverse Environmental Stresses," in *Yeast Stress Responses*, ed. S. Hohmann and W. Mager (Berlin: Springer Verlag, 2003), 13 (emphasis added).

3. Ibid., 15.

4. A. Gasch et al., "Genomic Expression Programs in the Response of Yeast Cells to Environmental Changes," *Molecular Biology of the Cell* 11 (2000): 4241–57.

5. D. Rangel, "Stress Induced Cross-Protection against Environmental Challenges on Prokaryotic and Eukaryotic Microbes," *World Journal of Microbiology and Biotechnology* 27 (2010): 1281–96.

6. J. I. Murray et al., "Diverse and Specific Gene Expression Responses to Stresses in Cultured Human Cells," *Molecular Biology of the Cell* 15 (2004): 2361–74.

7. F. Pampaloni et al., "The Third Dimension Bridges the Gap between Cell Culture and Live Tissue," *Nature Reviews Molecular Cell Biology* 8 (2007): 839–45.

8. C. McHale et al., "Toxicogenomic Profiling of Chemically Exposed Humans in Risk Assessment," *Mutation Research* 705 (2010): 172–83.

9. J. Goldstone et al., "The Chemical Defensome: Environmental Sensing and Response Genes in the *Strongylocentrotus purpuratus* Genome," *Developmental Biology* 300 (2006): 366.

10. J. Goldstone, "Environmental Sensing and Response Genes in Cnidaria: The Chemical Defensome in the Sea Anemone *Nematostella vectensis*," *Cell Biology and Toxicology* 24 (2008): 483–502; Goldstone et al., "Chemical Defensome."

11. M. Berenbaum, "Postgenomic Chemical Ecology: From Genetic Code to Ecological Interactions," *Journal of Chemical Ecology* 28 (2002): 875.

12. For information on toxicogenomics, see E. Hubal et al., "Exposure Science and the U.S. EPA National Center for Computational Toxicology," *Journal of Exposure Science and Epidemiology* 20 (2010): 231–36; "Toxicogenomics," *International Programme on Chemical Safety*, n.d., accessed August 2011, http://www.who.int/ipcs/methods/toxicogenomics/en/index.html; and E. Monosson, "Chemical Mixtures: Considering the Evolution of Toxicology and Chemical Assessment," *Environmental Health Perspectives* 113 (2004): 383–90.

13. Goldstone, "Environmental Sensing and Response Genes in Cnidaria." Goldstone and colleagues have yet to plumb the defensomes of species that parted ways more than 550 million years ago, before the divergence of cnidaria (corals, sea anemones, and jellyfish) from other bilaterians (protostomes and deuterostomes).

14. Although Goldstone's work has focused primarily on deuterostomes, he surmises based on his work with the sea anemone—a species close to the deuterostome-protostome split—that the defensome of other animals like the lophotrochozoans (non-molting protostomes animals including mollusks, segmented worms, and nematodes) would similarly constitute 2%–3% of the genome. J. Goldstone, e-mail communication, March 23, 2011.

15. D. Epel, "Protection of DNA during Early Development: Adaptations and Evolutionary Consequences," *Evolution and Development* 5 (2003): 83–88.

16. A. Hamdoun and D. Epel, "Embryo Stability and Vulnerability in an Always Changing World," 104 (2007): 1745.

17. Ibid.; Epel, "Protection of DNA during Early Development."

18. Hamdoun and Epel, "Embryo Stability and Vulnerability," 1745.

19. A. Hamdoun, e-mail communication, April 5, 2011.

20. Hamdoun and Epel, "Embryo Stability and Vulnerability."

21. A. Bird, "Perceptions of Epigenetics," *Nature* 447 (2007): 398.

22. Hamdoun and Epel, "Embryo Stability and Vulnerability," 1749. For more on epigenetics, see the following October 29, 2010, issue of the journal *Science*, with a special "Section on Epigenetics," accessed June 2011, http://www.sciencemag.org/content/330/6004.toc.

23. C. Guerrero-Bosagna et al., "Epigenetic Transgenerational Actions of Vinclozolin on Promoter Regions of the Sperm Epigenome," *PloS ONE* 5 (2010): e13100. For a review, see M. K. Skinner, M. Manikkam, and C. Guerrero-Bosagna, "Epigenetic Transgenerational Actions of Endocrine Disruptors," *Reproductive Toxicology* 31 (2010): 337–43.

24. See M. Mattson and E. Calabrese, eds., *Hormesis: A Revolution in Biology, Toxicology and Medicine* (Totowa, NJ: Humana Press, 2009).

25. E. J. Calabrese, "Hormesis: Why It Is Important to Toxicology and Toxicologists," *Environmental Toxicology and Chemistry* 27 (2008): 1451–74, p. 1451.

26. Ibid., 1458.

27. L. M. Gerber, G. Williams, and S. Gray, "The Nutrient-Toxin Dosage Continuum in Human Evolution and Modern Health," *Quarterly of Biology* 74 (1999): 273–89, p. 285.

28. G. Heinz et al., "Enhanced Reproduction in Mallards Fed a Low Level of Methylmercury: An Apparent Case of Hormesis," *Environmental Toxicology and Chemistry* 29 (2010): 650–53. Tempering their conclusions, the authors note that more work is needed to confirm these results and rule out, for example, any beneficial effects of mercury on possible underlying infections that may have affected both controls and treated animals. See J. Raloff, "Fowl Surprise! Methylmercury Improves Hatching Rate," *Science News*, March 5, 2010, accessed August 2011, http://www.sciencenews.org/view/generic/id/56954/title/Science_%2B_the_Public__Fowl_surprise!_Methyl mercury_improves_hatching_rate.

29. Gerber, Williams, and Gray, "Nutrient-Toxin Dosage Continuum."

30. A. Kashiwagi et al., "Adaptive Response of a Gene Network to Environmental Changes by Fitness-Induced Attractor Selection," *PloS ONE* 1 (2006): e49.

31. S. Huang, e-mail communication, March 22, 2011.

32. As Huang explains, "We cannot speak of 'real energy' since life itself is an island far away from thermodynamic equilibrium where the classical laws

of physics related to energetic operates. This term 'quasi-potential energy' is generally accepted in non-equilibrium thermodynamics." Ibid.

33. Ibid.

34. Kashiwagi et al., "Adaptive Response of a Gene Network."

35. Ibid., e49.

36. Huang, e-mail communication, March 22, 2011.

37. Kashiwagi et al., "Adaptive Response of a Gene Network," e49.

38. Quotation from Huang, e-mail communication, March 22, 2011. See also S. Huang, "Reprogramming Cell Fates: Reconciling Rarity with Robustness," *BioEssays* 31 (2009): 546–60.

Chapter 9

1. A. Fleming, "Penicillin: Nobel Lecture," Nobelprize.org, December 11, 1945, accessed August 2011, http://nobelprize.org/nobel_prizes/medicine/laureates/1945/fleming-lecture.pdf.

2. S. Palumbi, "Better Evolution through Chemistry: Rapid Evolution Driven by Human Changes to the Chemical Environment," *Chemical Evolution II: From the Origins of Life to Modern Society*, ACS Symposium Series 1025 (2009): 333–43.

3. This example has been the subject of ongoing research and controversy, particularly since the publication of *Melanism: Evolution in Action* by Michael Majerus (Oxford: Oxford University Press, 1998). While Majerus did not question the conclusion of rapid evolution by natural selection, he brought attention to the scientific methodology supporting the claim that predation by birds (on the more visible moths) was the primary driving force. Lacking any long-term observations of predation by birds, Henry Bernard Kettlewell, a British geneticist and physician, devised a series of experiments involving both dark and light moths, insect-eating birds and soot-impacted forest enclosures, and enclosures free from industrial influences. Majerus had reviewed several concerns about Kettlewell's experimental design, revealing some common pitfalls associated with recreating enclosed and controlled versions of field conditions. An apparently misunderstood review of the book published in the journal *Nature* followed, providing an unprecedented opportunity for creationists to challenge the scientific basis for evolution. The controversy eventually resulted in the removal of the iconic photos from one of the most widely used introductory biology texts, replaced instead with images of beaks from Darwin's finches. Subsequent studies, including those by Majerus, have returned the peppered moths (and their susceptibility to predation by birds) to their role as prime examples of directional selection. This is recounted in J. de Roode, "The Moths of War," *New Scientist*, December 8, 2007, accessed August 2011, http://leebor2.741.com/moth.html. For Majerus's own account, see "The Peppered Moth: The Proof of Darwinian

Evolution," n.d., accessed August 2011, http://www.gen.cam.ac.uk/research /personal/majerus/Swedentalk220807.pdf. See also L. Cook and J. Turner, "Decline of Melanism in Two British Moths: Spatial, Temporal and Inter-Specific Variation," *Heredity* 101 (2008): 483–89.

4. For a review, see A. Hendry et al., "Evolutionary Principles and Their Practical Application," *Evolutionary Applications* 4 (2011): 159–83.

5. S. J. Gould and N. Eldredge, "Punctuated Equilibrium Comes of Age," *Nature* 366 (1993): 223. These ideas were first presented in N. Eldredge and S. J. Gould, "Punctuated Equilibria: An Alternative to Phyletic Gradualism," in *Models in Paleobiology*, ed. T. Schopf (San Francisco: Freeman, Cooper, 1972), 82–115.

6. Gould and Eldredge, "Punctuated Equilibrium Comes of Age"; Gene Hunt, "Evolution in Fossil Lineages: Paleontology and the Origin of Species," *American Naturalist* 176 (2010): S61–S76.

7. P. R. Grant and B. R. Grant, "Predicting Microevolutionary Response to Directional Selection on Heritable Variation," *Evolution* 49 (1995): 241–51; P. R. Grant and B. R. Grant, "Unpredictable Evolution in a 30-Year Study of Darwin's Finches," *Science* 296 (2002): 707–11.

8. Grant and Grant, "Unpredictable Evolution in a 30-Year Study of Darwin's Finches," 709.

9. Ibid., 710.

10. A. Hendry and M. Kinnison, "The Pace of Modern Life: Measuring Rates of Contemporary Microevolution," *Evolution* 53 (1999): 1638.

11. Ibid., 1650.

12. Reviewed in R. Russell et al., "The Evolution of New Enzyme Function: Lessons from Xenobiotic Metabolizing Bacteria versus Insecticide-Resistant Insects," *Evolutionary Applications* 4 (2011): 225–48.

13. Ibid.

14. R. Barrett and D. Schluter, "Adaptation from Standing Genetic Variation," *Trends in Ecology and Evolution* 23 (2008): 38–44; T. G. Wilson, "Drosophila: Sentinels of Environmental Toxicants," *Integrative and Comparative Biology* 45 (2005): 127–36; T. Karasov, P. Messer, and D. Petrov, "Evidence That Adaptation in Drosophila Is Not Limited by Mutation at Single Sites," *PLoS Genetics* 6 (2010): e1000924.

15. For a review, see Barrett and Schluter "Adaptation from Standing Genetic Variation"; and J. Stapley et al., "Adaptation Genomics: The Next Generation," *Trends in Ecology and Evolution* 25 (2010): 705–12.

16. T. G. Wilson, "Drosophila: Sentinels of Environmental Toxicants," *Integrative and Comparative Biology* 45 (2005): 127–36; Stapley et al., "Adaptation Genomics."

17. C. Linnen et al., "On the Origin and Spread of an Adaptive Allele in Deer Mice," *Science* 325 (2009): 1095–98.

18. For a review, see D. Nacci, D. Champlin, and S. Jayaraman,

"Adaptation of the Estuarine Fish *Fundulus heteroclitus* (Atlantic Killifish) to Polychlorinated Biphenyls (PCBs)," *Estuaries and Coasts* 33 (2010): 853–64.

19. This is a complex issue that includes fishing, predation, and toxicity. The following study teases apart all these factors over the past century: P. Cook et al., "Effects of Aryl Hydrocarbon Receptor-Mediated Early Life Stage Toxicity on Lake Trout Populations in Lake Ontario during the 20th Century," *Environmental Science and Technology* 37 (2003): 3864–77.

20. This also brings to light the far-flung consequences of resistance in fish. By accumulating but not succumbing to toxic chemicals, they become a conduit for passing chemicals along to more sensitive species. See S. Bursian et al., "Dietary Exposure of Mink (*Mustela vison*) to Fish from the Housatonic River, Berkshire County, Massachusetts, USA: Effects on Organ Weights and Histology and Hepatic Concentrations of Polychlorinated Biphenyls and 2,3,7,8-Tetrachlorodibenzo-p-dioxin Toxic," *Environmental Toxicology and Chemistry* 25 (2006): 1541–50.

21. L. Wackett, "Questioning Our Perceptions about Evolution of Biodegradative Enzymes," *Current Opinion in Microbiology* 12 (2009): 244–51.

22. Some of the first studies of dioxin-resistant fish came out of Keith Cooper's laboratory. See R. Prince and K. R. Cooper, "Comparisons of the Effects of 2,3,7,8-Tetrachlorodibenzo-p-dioxin on Chemically Impacted and Nonimpacted Subpopulations of *Fundulus heteroclitus*: I. TCDD Toxicity," *Environmental Toxicology and Chemistry* 14 (1995): 579–87; and R. Prince and K. R. Cooper, "Comparisons of the Effects of 2,3,7,8-Tetrachlorodibenzo-p-dioxin on Chemically Impacted and Nonimpacted Subpopulations of *Fundulus heteroclitus*: II. Metabolic Considerations," *Environmental Toxicology and Chemistry* 14 (1995): 589–95.

23. A. Elskus, E. Monosson, and D. S. Woltering, "Altered CYP1A Expression in *Fundulus heteroclitus* Adults and Larvae: A Sign of Pollutant Resistance?," *Aquatic Toxicology* 45 (1999): 99–113.

24. D. W. Whitman and A. Agrawal, "What Is Phenotypic Plasticity and Why Is It Important?," in *Phenotypic Plasticity of Insects: Mechanisms and Consequences*, ed. D. Whitman and T. Ananthakrishna (Enfield, NH: Science Publishers, 2009), 1.

25. M. Kinnison, e-mail communication, April 11, 2011.

26. D. Nacci et al., "Evolution of Tolerance to PCBs and Susceptibility to a Bacterial Pathogen (*Vibrio harveyi*) in Atlantic Killifish (*Fundulus heteroclitus*) from New Bedford (MA, USA) Harbor," *Environmental Pollution* 157 (2009): 857–64; Nacci, Champlin, and Jayaraman, "Adaptation of the Estuarine Fish *Fundulus heteroclitus* (Atlantic Killifish) to Polychlorinated Biphenyls (PCBs)."

27. By definition, a truly common garden would require that different populations inhabit the same tanks. This is not always practicable.

28. M. Kinnison, e-mail communication, April 11, 2011.

29. Hendry and Kinnison, "Pace of Modern Life."

30. See A. Whitehead et al., "Comparative Transcriptomics Implicates Mechanisms of Evolved Pollution Tolerance in a Killifish Population," *Molecular Ecology* 19 (2010): 5186–203.

31. Ibid.

32. X. Arzuaga and A. Elskus, "Polluted-Site Killifish (*Fundulus heteroclitus*) Embryos Are Resistant to Organic Pollutant-Mediated Induction of CYP1A Activity, Reactive Oxygen Species, and Heart Deformities," *Environmental Toxicology and Chemistry* 29 (2010): 676–82; M. Hahn, L. Allan, and D. Sherr, "Regulation of Constitutive and Inducible AHR Signaling: Complex Interactions Involving the AHR Repressor," *Biochemical Pharmacology* 77 (2009): 485–97.

33. I. Wirgin, "Resistance to Contaminants in North American Fish Populations," *Mutation Research* 552 (2004): 73–100.

34. The finding is in contrast to the amino acid substitutions, which more directly interfere with binding as observed in resistant mice and other mammals (discussed in chapter 7), which also *seem* to be incidental rather than a consequence of contemporary evolution in response to PCB or dioxin contamination.

35. I. Wirgin et al., "Mechanistic Basis of Resistance to PCBs in Atlantic Tomcod from the Hudson River," *Science* 331 (2011): 1324.

36. J. Bickham and M. Smolen, "Somatic and Heritable Effects of Environmental Genotoxins and the Emergence of Evolutionary Toxicology," *Environmental Health Perspectives* 102 (suppl. 2) (1994): 25.

37. J. W. Bickham, "The Four Cornerstones of Evolutionary Toxicology," *Ecotoxicology* 20 (2011): 497–502.

38. Hendry and Kinnison, "Pace of Modern Life."

39. D. Nacci, e-mail communication, April 11, 2011.

40. Joshua Mackie et al., "Loss of Evolutionary Resistance by the Oligochaete *Limnodrilus hoffmeisteri* to a Toxic Substance—Cost or Gene Flow?," *Evolution* 64 (2010): 152–65.

41. Similar to the AhR's earliest function discussed in chapter 7, the AChE removes chemical messengers that are necessary but toxic or disruptive if allowed to build up.

42. P. Labbé et al., "Forty Years of Erratic Insecticide Resistance Evolution in the Mosquito *Culex pipiens*," *PLoS Genetics* 3 (2007): e205.

43. Mackie et al., "Loss of Evolutionary Resistance"; D. Conover, "Nets versus Nature," *Nature* 450 (2005): 179–80; D. Conover and S. Munch, "Sustaining Fisheries Yields over Evolutionary Time Scales," *Science* 297 (2002): 94–96.

44. Reviewed in Russell et al., "Evolution of New Enzyme Function"; see also Barrett and Schluter, "Adaptation from Standing Genetic Variation."

45. Labbé et al., "Forty Years of Erratic Insecticide Resistance Evolution," 2190.

46. Ibid.

47. Ibid., 2196.

48. For a review, see M. Medina, J. Correa, and C. Barata, "Micro-Evolution Due to Pollution: Possible Consequences for Ecosystem Responses to Toxic Stress," *Chemosphere* 67 (2007): 2105–14.

49. A. Hendry et al., "Evolutionary Principles and Their Practical Application."

50. Hendry and Kinnison, "Pace of Modern Life," 1650.

Chapter 10

1. A. Barnosky et al., "Has the Earth's Sixth Mass Extinction Already Arrived?," *Nature* 471 (2011): 56.

2. Ibid.

3. B. Gill et al., "Geochemical Evidence for Widespread Euxinia in the Later Cambrian Ocean," *Nature* 469 (2011): 80–83.

4. P. Vitousek et al., "Human Domination of Earth's Ecosystems," *Science* 277 (1997): 494.

5. Ibid., 498.

6. See "Materials," *Plastics News*, October 30, 2009, accessed July 2011, http://plasticsnews.com/fyi-charts/materials.html?id=17004.

7. J. Raloff, "50 Million Chemicals and Counting," *Science News*, September 8, 2009, accessed August 2011, http://www.usnews.com/science/articles/2009/09/08/50-million-chemicals-and-counting_print.html.

8. L. Wackett, "Questioning Our Perceptions about Evolution of Biodegradative Enzymes," *Current Opinion in Microbiology* 12 (2009): 244–51.

9. T. Hartung, "From Alternative Methods to a New Toxicology," *European Journal of Pharmaceutics and Biopharmaceutics* 77 (2011): 338–49.

10. C. de Wit, D. Herzke, and K. Vorkamp, "Brominated Flame Retardants in the Arctic Environment—Trends and New Candidates," *Science of the Total Environment* 408 (2010): 2885–2918; A. Derocher et al., "Contaminants in Svalbard Polar Bear Samples Archived since 1967 and Possible Population Level Effects," *Science of the Total Environment* 301 (2003): 163–74; "National Report on Human Exposure to Environmental Chemicals," *Centers for Disease Control and Prevention*, 2011, accessed June 2011, http://www.cdc.gov/exposurereport/.

11. For more, see M. Wilson and M. Schwarzman, "Toward a New U.S. Chemicals Policy: Rebuilding the Foundation to Advance New Science, Green Chemistry, and Environmental Health," *Environmental Health Perspectives* 117 (2009): 1202–9.

12. "Assessing Chemical Risk: Societies Offer Expertise," *Science* 331 (2011): 1136.

13. See "National Health and Nutrition Examination Survey," *Centers for Disease Control and Prevention*, 2011, accessed June 2011, http://www.cdc .gov/nchs/nhanes.htm.

14. D. Tillitt et al., "Atrazine Reduces Reproduction in Fathead Minnow (*Pimephales promelas*)," *Aquatic Toxicology* 99 (2010): 149–59; S. Chen and M. DeWitt, "Murky Water: Science, Money, and the Battle over Atrazine," *Berkeley Science Review* 20 (2011): 44–52; "Atrazine Updates," *U.S. Environmental Protection Agency*, June 2011, accessed August 2011, http://www.epa .gov/opp00001/reregistration/atrazine/atrazine_update.htm; "Atrazine Reevaluation: Introduction and Status," *U.S. Environmental Protection Agency*, September 14–17, 2010, accessed August 2011, http://www.regulations.gov /#!documentDetail;D=EPA-HQ-OPP-2010-0481-0048;oldLink=false.

15. K. Harley et al., "PBDE Concentrations in Women's Serum and Fecundability," *Environmental Health Perspectives* 118 (2010): 699–704.

16. S.-K. Kim et al., "Distribution of Perfluorochemicals between Sera and Milk from the Same Mothers and Implications for Prenatal and Postnatal Exposures," *Environmental Pollution* 159 (2011): 169–74; B. Kelly et al., "Perfluoroalkyl Contaminants in an Arctic Marine Food Web: Trophic Magnification and Wildlife Exposure," *Environmental Science and Technology* 43 (2009): 4037–43. See also "Perfluoroalkyls," *Agency for Toxic Substances and Disease Registry*, May 2009, accessed August 2011, http://www.atsdr.cdc.gov /toxprofiles/tp.asp?id=1117&tid=237.

17. D. Krewski et al., *Toxicity Testing in the 21st Century: A Vision and a Strategy* (Washington, DC: National Academies Press, 2007), accessed August 2011, http://www.nap.edu/catalog.php?record_id=11970#toc.

18. M. Andersen and D. Krewski, "The Vision of Toxicity Testing in the 21st Century: Moving from Discussion to Action," *Toxicological Sciences* 117 (2010): 18.

19. Ibid., 23.

20. Ibid., 17.

21. Ibid., 23.

22. Krewski et al., *Toxicity Testing in the 21st Century*.

23. T. Hartung and M. McBride, "Food for Thought: On Mapping the Human Toxome," *ALTEX* 28 (2011): 83–93.

24. Ibid., 85.

25. R. Carson, *Silent Spring* (New York: Houghton Mifflin, 1962).

26. J. W. Bickham, "The Four Cornerstones of Evolutionary Toxicology," *Ecotoxicology* 20 (2011): 498.

27. For the text of Feynman's speech, which is credited with spurring on the development of nanotechnology, see "There's Plenty of Room at the Bottom," December 1959, accessed October 2011, http://www.physics.unc.edu

/~falvo/Phys53_Spring11/Reading_Assignments/New_Papers/Feynman
_Plenty_of_Room.pdf.

28. P. Anastas and N. Eghbali, "Green Chemistry: Principles and Practice," *Chemical Society Reviews* 39 (2010): 301–12.

Appendix

1. Review S. Safe, "Toxicology, Structure-Function Relationship, and Human and Environmental Health Impacts of Polychlorinated Biphenyls: Progress and Problems," *Environmental Health* 100 (1992): 259–68.

2. H.-Y. Park et al., "Neurodevelopmental Toxicity of Prenatal Polychlorinated Biphenyls (PCBs) by Chemical Structure and Activity: A Birth Cohort Study," *Environmental Health* 9 (2010): 9–51; S. White and L. Birnbaum, "An Overview of the Effects of Dioxins and Dioxin-Like Compounds on Vertebrates, as Documented in Human and Ecological Epidemiology," *Journal of Environmental Science and Health: Part C, Environmental Carcinogenesis and Ecotoxicology Reviews* 27 (2009): 197–211.

3. K. Breivik et al., "Towards a Global Historical Emission Inventory for Selected PCB Congeners—a Mass Balance Approach 3. An Update," *Science of the Total Environment* 377 (2007): 296–307.

4. C. Bogdal et al., "Release of Legacy Pollutants from Melting Glaciers: Model Evidence and Conceptual Understanding," *Environmental Science and Technology* 44 (2010): 4063–69.

5. See M. Harada, "Minamata Disease: Methylmercury Poisoning in Japan Caused by Environmental Pollution," *Critical Reviews in Toxicology* 25 (1995): 1–24; P. Grandjean et al., "Adverse Effects of Methylmercury: Environmental Health Research Implications," *Environmental Health Perspectives* 118 (2010): 1137–45.

6. See N. Selin, "Global Biogeochemical Cycling of Mercury: A Review," *Annual Review of Environment and Resources* 34 (2009): 43–63.

7. A recent study in mallards reported enhanced reproduction. See G. H. Heinz et al., "Enhanced Reproduction in Mallards Fed a Low Level of Methylmercury: An Apparent Case of Hormesis," *Environmental Toxicology and Chemistry* 29 (2010): 650–53. But determining whether increases in reproductive end points (including growth, survival, and fecundity) are beneficial effects is complex and requires greater understanding of the immediate and long-term consequences of enhanced reproduction.

8. C. Vetriani et al., "Mercury Adaptation among Bacteria from a Deep-Sea Hydrothermal Vent," *Applied Environmental Microbiology* 71 (2005): 220–26; M. Mulvey et al., "Genetic and Demographic Responses of Mosquitofish (*Gambusia holbrooki* Girard 1859) Populations Stressed by Mercury," *Environmental Toxicology and Chemistry* 14 (1995): 1411–18.

9. For a review of mercury and its toxicity, see T. Clarkson and L. Magos, "The Toxicology of Mercury and Its Chemical Compounds," *Critical Reviews in*

Toxicology 36 (2006): 609–62; T. Clarkson, "Health Effects Metals: A Role for Evolution?," *Environmental Health Perspectives* 103 (suppl.) 1 (1995): 9–12.

10. E . Pacyna et al., "Global Anthropogenic Mercury Emission Inventory for 2000," *Atmospheric Environment* 40 (2006): 4048–63; Selin, "Global Biogeochemical Cycling of Mercury."

11. L.-Q. Xu et al., "A 700-Year Record of Mercury in Avian Eggshells of Guangjin Island, South China Sea," *Environmental Pollution* 159 (2011): 889–96.

12. Selin, "Global Biogeochemical Cycling of Mercury"; Pacyna et al., "Global Anthropogenic Mercury Emission Inventory for 2000."

13. Ibid., 55.

14. L. N. Plummer and E. Busenberg, "Chlorofluorocarbons Background," *USGS*, November 27, 2009, accessed August 2011, http://water .usgs.gov/lab/chlorofluorocarbons/background/.

15. Ibid.

16. U.S. Food and Drug Administration, "Epinephrine CFC Metered-Dose Inhalers," n.d., accessed August 2011, http://www.fda.gov/downloads /Drugs/DrugSafety/InformationbyDrugClass/UCM182381.pdf; UNEP Technology and Economic Assessment Panel, *Montreal Protocol on Substances That Deplete the Ozone Layer* (Nairobi, Kenya: United Nations Environment Programme, 2010), accessed August 2011, http://ozone.unep.org/teap /Reports/TEAP_Reports/teap-2010-progress-report-volume2-May2010.pdf.

17. World Meteorological Organization, "Record Stratospheric Ozone Loss in the Arctic in the Spring of 2011," press release, April 5, 2011, accessed August 2011, http://ozone.unep.org/Publications/912_en.pdf.

18. J. McLachlan, "Environmental Signaling: What Embryos and Evolution Teach Us about Endocrine Disrupting Chemicals," *Endocrine Reviews* 22 (2001): 335.

19. Ibid.

20. Ibid., 336.

21. T. Colborn, D. Dumanowski, and P. Meyers, *Our Stolen Future* (New York: Penguin, 1996); S. Krimsky, *Hormonal Chaos* (Baltimore: Johns Hopkins University Press, 2002); Endocrine Disruptor Screening and Testing Advisory Committee, *U.S. Environmental Protection Agency*, August 11, 2011, accessed August 2011, http://www.epa.gov/endo/pubs/edspoverview/edstac .htm.

22. S. Holley, "Nano Revolution—Big Impact: How Emerging Nanotechnologies Will Change the Future of Education and Industry in America," *Journal of Technology Studies* 35 (2009), accessed August 2011, http:// scholar.lib.vt.edu/ejournals/JOTS/v35/v35n1/pdf/holley.pdf. See also a presentation by the National Institute for Occupational Health and Safety, "Preventing Adverse Health Effects from Nanotechnology," April 15, 2010, accessed August 2011, http://www.cdc.gov/about/grand-rounds/archives /2010/download/GR-041510.pdf.

23. J. Clarence Davies, *EPA and Nanotechnology: Oversight for the 21st Century* (Washington, DC: Woodrow Wilson International Center for Scholars, Project on Emerging Technologies, 2007), accessed October 25, 2011, http://www.nanotechproject.org/process/assets/files/2698/197_nanoepa_pen9.pdf.

24. Davis et al., *Nanomaterial Case Studies: Nanoscale Titanium Dioxide in Water Treatment and in Topical Sunscreen* (Washington, DC: U.S. Environmental Protection Agency, 2010), accessed August 2011, http://cfpub.epa.gov/ncea/cfm/recordisplay.cfm?deid=230972.

25. Richard Feynman, "There Is Plenty of Room at the Bottom," December 1959, accessed August 2011, http://www.physics.unc.edu/~falvo/Phys53_Spring11/Reading_Assignments/New_Papers/Feynman_Plenty_of_Room.pdf; T. Yih and V. Moudgil, "Nanotechnology Comes of Age to Trigger the Third Industrial Revolution," *Nanomedicine: Nanotechnology, Biology, and Medicine* 3 (2007): 245.

26. See the Project on Emerging Nanotechnologies, sponsored by the Woodrow Wilson International Center for Scholars and the Pew Charitable Trusts, accessed August 2011, http://www.nanotechproject.org/about/mission/; "List of Nanotechnology Organizations," *Wikipedia*, n.d., accessed September 30, 2011, http://en.wikipedia.org/wiki/List_of_nanotechnology_organizations.

27. Günter Oberdörster et al., "Principles for Characterizing the Potential Human Health Effects from Exposure to Nanomaterials: Elements of a Screening Strategy," *Particle and Fibre Toxicology* 2 (2005): 8.

28. See A. Nel et al., "Toxic Potential of Materials at the Nanolevel," *Science* 311 (2006): 622–27.

29. A feature article in *Science News* from February 28, 2009, reported a number of research efforts linking mitochondrial failure or weakening to neurodegenerative and aging-related conditions and diseases. See L. Bell, "Mitochondria Gone Bad," *Science News* 175 (2009): 20–23. See also K. Unfried et al., "Cellular Responses to Nanoparticles: Target Structures and Mechanisms," *Nanotoxicology* 1 (2007): 57–71. Nanoparticles may even cause toxicity indirectly, without crossing the cell membrane, perhaps through interactions with some of the most ancient and universally shared receptors, including the purine receptors mentioned in chapter 7. See F. Marano et al., "Nanoparticles: Molecular Targets and Cell Signaling," *Archives of Toxicology* 85 (2011): 733–41; L. Fei and S. Perrett, "Effect of Nanoparticles on Protein Folding and Fibrillogenesis," *International Journal of Molecular Sciences* 10 (2009): 646–55.

SELECTED BIBLIOGRAPHY

Barnosky, A., N. Matzke, S. Tomiya, G. O. U. Wogan, B. Swartz, T. B. Quental, C. Marshall, et al. "Has the Earth's Sixth Mass Extinction Already Arrived?" *Nature* 471 (2011): 51–57.

Barrett, R., and D. Schluter. "Adaptation from Standing Genetic Variation." *Trends in Ecology and Evolution* 23 (2008): 38–44.

Belyi, V., P. Ak, E. Markert, H. Wang, W. Hu, A. Puzio-Kuter, and A. J. Levine. "The Origins and Evolution of the p53 Family of Genes." *Cold Spring Harbor Perspectives in Biology* 2 (2010): a001198.

Berenbaum, M. "Postgenomic Chemical Ecology: From Genetic Code to Ecological Interactions." *Journal of Chemical Ecology* 28 (2002): 873–96.

Bickham, J. "The Four Cornerstones of Evolutionary Toxicology." *Ecotoxicology* 20 (2011): 497–502.

Bickham, J., and M. Smolen. "Somatic and Heritable Effects of Environmental Genotoxins and the Emergence of Evolutionary Toxicology." *Environmental Health Perspectives* 102 (suppl. 2) (1994): 25–28.

Calabrese, E. "Hormesis: Why It Is Important to Toxicology and Toxicologists." *Environmental Toxicology and Chemistry* 27 (2008): 1451–74.

Calabrese, E., and L. Baldwin. "Chemical Hormesis: Its Historical Foundations as a Biological Hypothesis." *Human and Experimental Toxicology* 19 (2000): 2–31.

Carroll, S. *Endless Forms Most Beautiful*. New York: Norton, 2005.

Castresana, J., and M. Seraste. "Evolution of Energetic Metabolism: The Respiration-Early Hypothesis." *Trends in Biological Sciences* 20 (1995): 443–48.

Cockell, C., and A. Blaustein, eds. *Ecosystems, Evolution, and Ultraviolet Radiation*. New York: Springer, 2001.

Colborn, T., D. Dumanowski, and P. Meyers. *Our Stolen Future*. New York: Penguin, 1996.

Conover, D., and S. Munch. "Sustaining Fisheries Yields over Evolutionary Time Scales." *Science* 297 (2002): 94–96.

Costantini, D. "Oxidative Stress in Ecology and Evolution: Lessons from Avian Studies." *Ecology Letters* 11 (2008): 1238–51.

Coyle, P., J. Philcox, L. Carey, and A. Rofe. "Metallothionein: The Multipurpose Protein." *Cellular and Molecular Life Sciences* 59 (2002): 627–47.

Crespi, B., and K. Summers. "Evolutionary Biology of Cancer." *Trends in Ecology and Evolution* 20 (2005): 545–52.

Dawkins, R. *The Ancestor's Tale: A Pilgrimage to the Dawn of Evolution*. New York: Houghton Mifflin, 2004.

de Roode, J. "The Moths of War." *New Scientist*, December 8, 2007. Accessed August 2011. http://leebor2.741.com/moth.html.

Dobzhansky, T. "Nothing in Biology Makes Sense except in the Light of Evolution." *American Biology Teacher* 35 (1973): 125–29.

Domazet-Loso, T., and D. Tautz. "Phylostratigraphic Tracking of Cancer Genes Suggests a Link to the Emergence of Multicellularity in Metazoa." *BioMed Central Biology* 8 (2010): 66.

Duke, J. "Medicinal Plants and the Pharmaceutical Industry." In *New Crops*, edited by J. Janick and J. E. Simon, 664–69. New York: Wiley, 1993.

Dupont, C., A. Butcher, R. Valas, P. Bourne, and G. Caetano-Anolles. "History of Biological Metal Utilization Inferred through Phylogenomic Analysis of Protein Structures." *Proceedings of the National Academy of Sciences* 107 (2010): 10567–72.

Dupont, C., and G. Caetano-Anolles. "Reply to Mulkidjanian and Galperin: Zn May Have Constrained Evolution during the Proterozoic but Not the Archean." *Proceedings of the National Academy of Sciences* 107 (2010): E138.

Escriva, H., F. Delaunay, and V. Laudet. "Ligand Binding and Nuclear Receptor Evolution." *BioEssays* 22 (2000): 717–27.

Estabrook, R. "A Passion for P450s (Remembrances of the Early History of Research on Cytochrome P450)." *Drug Metabolism and Disposition* 31 (2003): 1461–73.

Feyereisen, R. "Evolution of Insect P450." *Biochemical Society Transactions* 34 (2006): 1252–55.

Forgash, J. "History, Evolution and Consequences of Insecticide Resistance." *Pesticide Biochemistry and Biophysiology* 22 (1984): 178–86.

Forterre, P. "The Search for LUCA." *1997 Workshop Proceedings*. Accessed August 2011. http://translate.google.com/translate?hl=en&sl=fr&u=http://www-archbac.u-psud.fr/meetings/lestreilles/treilles_frm.html&ei=ZzHXS 87dOIaglAetkriABA&sa=X&oi=translate&ct=result&resnum=2&ved =0CAwQ7gEwAQ&prev=/search%3Fq%3Dbaptizing%2BLUCA%2 Bouzounis%26hl%3Den.

Fridovich, I. "Oxygen Toxicity: A Radical Explanation." *Journal of Experimental Biology* 201 (1998): 1203–9.

Gasch, A., P. Spellman, C. Kao, O. Carmel-Harel, M. Eisen, G. Storz, D. Botstein, et al. "Genomic Expression Programs in the Response of Yeast Cells to Environmental Changes." *Molecular Biology of the Cell* 11 (2000): 4241–57.

Geeta, E., and J. Thornton. "Evolution of Steroid Receptors from an Estrogen-Sensitive Ancestral Receptor." *Molecular and Cellular Endocrinology* 334 (2011): 31–38.

Gerber, L., G. Williams, and S. Gray. "The Nutrient-Toxin Dosage Continuum in

Human Evolution and Modern Health." *Quarterly Review of Biology* 74 (1999): 273–89.

Gluckman, P., A. Beedle, and M. Hanson, eds. *Principles of Evolutionary Medicine*. Oxford: Oxford University Press, 2010.

Goldstone, J. "Environmental Sensing and Response Genes in Cnidaria: The Chemical Defensome in the Sea Anemone *Nematostella vectensis*." *Cell Biology and Toxicology* 24 (2008): 483–502.

Goldstone, J., A. Hamdoun, B. J. Cole, M. Howard-Ashby, D. W. Nebert, M. Scally, M. Dean, et al. "The Chemical Defensome: Environmental Sensing and Response Genes in the *Strongylocentrotus purpuratus* Genome." *Developmental Biology* 300 (2006): 366–84.

Gould, S., and N. Eldredge. "Punctuated Equilibrium Comes of Age." *Nature* 366 (1993): 223–27.

Grant, P., and B. Grant. "Predicting Microevolutionary Response to Directional Selection on Heritable Variation." *Evolution* 49 (1995): 241–51.

———. "Unpredictable Evolution in a 30–Year Study of Darwin's Finches." *Science* 296 (2002): 707–11.

Greaves, M. *Cancer: The Evolutionary Legacy*. Oxford: Oxford University Press, 2000.

Grosberg, R., and R. Strathmann. "The Evolution of Multicellularity: A Minor Major Transition?" *Annual Review of Ecology, Evolution, and Systematics* 38 (2007): 621–54.

Gunsolus, J. "Herbicide Resistant Weeds." Regents of the University of Minnesota, 2002. Accessed August 2011. http://www.extension.umn.edu /distribution/cropsystems/dc6077.html.

Gutierrez, J., F. Amaro, and A. Martin-Gonzalez. "From Heavy Metal-Binders to Biosensors: Ciliate Metallothioneins Discussed." *BioEssays* 31 (2009): 805–16.

Hahn, M. "Aryl Hydrocarbon Receptors: Diversity and Evolution." *Chemico-Biological Interactions* 141 (2002): 131–60.

Hahn M., and S. Karchner. "Structural and Functional Diversification of AHRs during Metazoan Evolution," in *The AH Receptor in Biology and Toxicology*, edited by R. Pohjanvirta. New York: Wiley, forthcoming.

Hamdoun, A., and D. Epel. "Embryo Stability and Vulnerability in an Always Changing World." *Proceedings of the National Academy of Sciences* 104 (2007): 1745–50.

Hanahan, D., and R. Weinberg. "The Hallmarks of Cancer." *Cell* 100 (2000): 57–70.

Hartung, T. "Toxicology for the Twenty-First Century." *Nature* 460 (2009): 208–12.

———. "From Alternative Methods to a New Toxicology." *European Journal of Pharmaceutics and Biopharmaceutics* 77 (2011): 338–49.

Hartung, T., and M. McBride. "Food for Thought: On Mapping the Human Toxome." *ALTEX* 28 (2011): 83–93.

Hazen, R. "Evolution of Minerals." *Scientific American* 302 (2010): 58–65.

Hendry, A., and M. Kinnison. "The Pace of Modern Life: Measuring Rates of Contemporary Microevolution." *Evolution* 53 (1999): 1637–53.

Hendry, A., M. Kinnison, M. Heino, T. Day, T. Smith, G. Fitt, C. Bergstrom, et al. "Evolutionary Principles and Their Practical Application." *Evolutionary Applications* 4 (2011): 159–83.

Huang, S. "Reprogramming Cell Fates: Reconciling Rarity with Robustness." *BioEssays* 31 (2009): 546–60.

Janssens, T., D. Roelofs, and N. van Straalen. "Molecular Mechanisms of Heavy Metal Tolerance and Evolution in Insects." *Insect Science* 16 (2009): 3–18.

Junttila, M., and G. Evan. "p53—a Jack of All Trades but Master of None." *Nature Reviews Cancer* 9 (2009): 821–29.

Khakh, B., and G. Burnstock. "The Double Life of ATP." *Scientific American* 6 (2009): 84–92.

Kirschvink, J., and R. Kopp. "Palaeoproterozoic Ice Houses and the Evolution of Oxygen-Mediating Enzymes: The Case for a Late Origin of Photosystem II." *Philosophical Transactions of the Royal Society, Series B: Biological Sciences* 363 (2008): 2755–65.

Koonin, E. "Darwinian Evolution in Light of Genomics." *Nucleic Acids Research* 37 (2009): 1011–34.

Krewski, D., D. Acosta Jr., M. Andersen, H. Anderson, J. Bailar III, K. Boekelheide, R. Brent, et al. *Toxicity Testing in the 21st Century: A Vision and a Strategy*. Washington, DC: National Academies Press, 2007. Accessed August 2011. http://www.nap.edu/catalog.php?record_id=11970#toc.

Labbé, P., C. Berticat, A. Berthomieu, S. Unal, C. Bernard, M. Weill, and T. Lenormand. "Forty Years of Erratic Insecticide Resistance Evolution in the Mosquito *Culex pipiens*." *PLoS Genetics* 3 (2007): e205.

Lane, N. *Oxygen: The Molecule That Made the World*. Oxford: Oxford University Press, 2009.

Lucas-Lledó, J., and M. Lynch. "Evolution of Mutation Rates: Phylogenomic Analysis of the Photolyase/Cryptochrome Family." *Molecular Biology and Evolution* 26 (2009): 1143–53.

Lynch, M. "The Origins of Eukaryotic Gene Structure." *Molecular Biology and Evolution* 23 (2006): 450–68.

Ma, Q. "Xenobiotic-Activated Receptors: From Transcription to Drug Metabolism to Disease." *Chemical Research in Toxicology* 21 (2008): 1651–71.

Majerus, M. *Melanism: Evolution in Action*. Oxford: Oxford University Press, 1998.

Markov, G., and V. Laudet. "Origin and Evolution of the Ligand-Binding Ability of Nuclear Receptors." *Molecular and Cellular Endocrinology* 334 (2011): 21–30.

Mattson, M., and E. Calabrese, eds. *Hormesis: A Revolution in Biology, Toxicology and Medicine*. New York: Humana Press, 2009.

Medina, M., J. Correa, and C. Barata. "Micro-Evolution Due to Pollution: Possible Consequences for Ecosystem Responses to Toxic Stress." *Chemosphere* 67 (2007): 2105–14.

Metcalfe, N., and C. Alonso-Alvarez. "Oxidative Stress as a Life-History Constraint: The Role of Reactive Oxygen Species in Shaping Phenotypes from Conception to Death." *Functional Ecology* 24 (2010): 984–96.

Mulkidjanian, A., and M. Galperin. "On the Origin of Life in the Zinc World, 2: Validation of the Hypothesis on the Photosynthesizing Zinc Sulfide Edifices as Cradles of Life on Earth." *Biology Direct* 4 (2009): 27.

Nebert, D. W., and M. Z. Dieter. "The Evolution of Drug Metabolism." *Pharmacology* 61 (2000): 124–35.

Nelson, D. "The Cytochrome P450 Home Page." *Human Genomics* 4 (2009): 59–65.

Nielsen, F. "Evolutionary Events Culminating in Specific Minerals Becoming Essential for Life." *European Journal of Nutrition* 39 (2000): 62–66.

Ouzounis, C., V. Kunin, N. Darzentas, and L. Goldovsky. "A Minimal Estimate for the Gene Content of the Last Universal Common Ancestor—Exobiology from a Terrestrial Perspective." *Research in Microbiology* 157 (2006): 57–68.

Palumbi, S. "Better Evolution through Chemistry: Rapid Evolution Driven by Human Changes to the Chemical Environment." *Chemical Evolution II: From the Origins of Life to Modern Society*. ACS Symposium Series 1025 (2009): 333–43.

Poole, A. "My Name Is LUCA." *ActionBioscience*, September 2002. Accessed August 2011. http://www.actionbioscience.org/newfrontiers/poolepaper.html.

Ricardo, A., and J. Szostak. "Life on Earth." *Scientific American* 301 (2009): 54–61.

Rokas, A. "The Origins of Multicellularity and the Early History of the Genetic Toolkit for Animal Development." *Annual Review of Genetics* 42 (2008): 235–51.

Schuler, M. "P450s in Plant-Insect Interactions." *Biochimica et Biophysica Acta* 1814 (2011): 36–45.

Srivastava, M., O. Simakov, J. Chapman, B. Fahey, M. Gauthier, T. Mitros, G. Richards, et al. "The *Amphimedon queenslandica* Genome and the Evolution of Animal Complexity." *Nature* 466 (2010): 720–26.

Stapley, J., J. Reger, P. Feulner, C. Smadja, J. Galindo, R. Ekblom, C. Bennison, et al. "Adaptation Genomics: The Next Generation." *Trends in Ecology and Evolution* 25 (2010): 705–12.

Summers, A. "Damage Control: Regulating Defenses against Toxic Metals and Metalloids." *Current Opinion in Microbiology* 12 (2009): 138–44.

Thornton, J. "Nonmammalian Nuclear Receptors: Evolution and Endocrine Disruption." *Pure and Applied Chemistry* 75 (2003): 1827–39.

Valls, M., R. Bofill, R. Gonzalez-Duarte, P. Gonzalez-Duarte, M. Capdevila, and S. Atrian. "A New Insight in Metallothionein (MT) Classification and Evolution." *Journal of Biological Chemistry* 276 (2001): 32835–43.

van Straalen, N., and D. Roelofs. *Introduction to Ecological Genomics*. Oxford: Oxford University Press, 2006.

Wackett, L. "Questioning Our Perceptions about Evolution of Biodegradative Enzymes." *Current Opinion in Microbiology* 12 (2009): 244–51.

Williams, R., and J. Frausto da Silva. *The Chemistry of Evolution*. Amsterdam: Elsevier, 2006.

Wink, M. "Evolution of Secondary Metabolites from an Ecological and Molecular Phylogenetic Perspective." *Phytochemistry* 64 (2003): 3–19.

———. "Plant Secondary Metabolism: Diversity, Function and Its Evolution." *Natural Products Communications* 8 (2008): 1205–16.

Acetylcholine, 145–46
Acetylcholinesterase (AChE), 145
Adaptation, 30, 117, 124–25, 127, 133, 135, 140
Adenosine triphosphate (ATP), 42, 50, 103
Aerobic exercise, and oxygen stress, 44–45
Air pollutants, and ROS generation in lungs, 44, 46
Alleles, 145–46, 153
Allelochemicals, 88–89
Amphibian populations, 27–28
Anderson, Melvin, 155–56
Androgen receptors (AR), 102, 104, 107–8, 110
Antibiotics, 134
Antioxidants, 3, 42–43, 45–46
Apoptosis, 71, 76, 79, 121
Arctic, 164
Arthropods, terrestrial, 86
Aryl hydrocarbon receptors (AhR): activation of, in embryos sensitive to PCBs, 142; consequences of chemicals binding to, 143; CYP1A1 and, 94; dioxin affinity, 112–15; evolutionary origins of, 113–14; PCB cogeners and, 161–62; protein and ligand interactions involved in function of, 113; rapid evolution in response to local contamination, 114–15; role in absence of xenobiotic agonists in normal development, 144–45
Atlantic tomcod, 142–43
Atmospheric oxygen, 4, 92

Atrazine, 153–54
Attractor states, 127–30

Bacterial genome, 137
Belden, Lisa, 28–29
Benzo(a)pyrene (B[a]P), 29, 112
Berenbaum, May, 96, 98, 122–23
Beta-methylamino-l-alanine (BMAA), 88
Bickham, John, 133, 143, 157
Bioinformatics, defined, 20
Biological receptors, 102, 161
Biological toxins, 87–88
Bisphenol A (BPA), 152, 198n32
Bivalves, 78
Black swallowtails, 96–97
Blaustein, Andrew, 27–29
Blue mussels, 72–73
Burnstock, Geoff, 101, 103

Cadmium (Cd), 58, 62
Cadmium-binding proteins, 60
Cambialistic proteins, 56
Cancer, 66f, 67–68, 72–75
Cancer genome research, 75–76
Cannabinoid receptors, 102
Caretaker genes, 72, 75, 79–80
Carson, Rachel, 157
Catalase, 32f, 40–41, 42
Cellular repair, need and cost of, 75
Chemical Abstracts Service, 151–52
Chemical compounds, synthesis and breakdown of, 92
Chemical defense systems, 118
Chemical defensome, 122–24
Chemical/industrial revolution, 7
Chemical messengers, 100f, 103

Chemical mixtures, 8, 39, 162
Chemical revolution, nature's, 83–84
Chemicals: accumulation by species, 162; consumer-use, 110; current-use, 152; endogenous, 102; industrial, 107–8, 111, 152–53; intergenerational impacts of long-term exposure, 152–54; lipophilic, 106; modern, challenge ancient defenses, 148f; nano-formulations of, 169–72; synthetic, 151–52, 161–62. *See also* Toxic chemicals
Chemical-sensing receptors, 102
Chemical signals, 103
Chemical transmission, 101
Chlorofluorocarbons (CFCs), 25–26, 165–66
Choanozoans, 76
Chromium isotopes in BIF, 38
Chromophores, 16
Chytrid fungus, 29
Ciliates, 60
Climate change, and increased UVB exposure, 27
Cockell, Charles S., 15, 19
Colborn, Theo, 107–8
Coliform bacteria, 120
"Common garden" conditions, 141
Complex cells, 68–69
Complexity theory, 128
Constitutive receptors, 106
Contemporary/rapid evolution: in AhR in response to local contamination, 114–15; cancer cells, 75; costs of, 144–47; as defensive response to toxic chemicals, 143; determining a species' capacity for, 147; genetic mechanisms underlying, 137; meaning of term, 9–10; metal availability and, 62–64; observations of, in highly contaminated sites, 138; persistent exposure to toxins and, 134; phenotypic plasticity vs., 140; of pollution resistance in freshwater worms, 63; in response to synthetic chemicals, 152; standing genetic variation as key mechanism of, 63–64, 138, 147; in vertebrate populations, 136
Convergent evolution, 141–42
Copper-binding proteins, 56
Costantini, David, 45–46
Crespi, Barnard, 67, 75
Cryptochrome, 22
Cyanobacteria, 55, 61, 87–88, 189n10
CYPome, 84
Cytochrome P450 (CYP): chemical defensive system, 84, 91–95; clans, 93–95; classification, 93; CYP1A, 112, 139; CYP1A1, 93–94; CYP3A4, 97–98; CYP6 family, 95–97; CYP7, 95; CYP51, 92; as enzyme superfamily, 2, 57; evolutionary history of, 82f, 91; furanocoumarins and, 95–99; grapefruit juice and, 97–98; in metabolism and transformation of toxic chemicals, 90–91; relationship between dioxin, AhR, and, 112–13

Darwin's finches, 136
DDT, 108
Defense networks, 124–26
Defensive responses, 3
Defensome, 122–24
Deoxyribonucleic acid (DNA): binding region in nuclear receptors, 105–6; elements required for, 53; flexible regions and gene regulation, 19; induced adaptations, 125; kinking, 29; mutations, 15–16, 19, 73–74; noncoding, 20; nucleotide base pairs, 20; p53-mediated damage to, 81; repair systems, 3, 23–24, 28; ROS and integrity of, 73; turning on by nuclear receptors (NR), 104–7
Detoxification enzymes. *See* Cytochrome P450 (CYP)
Deuterostomes, 86, 93–94

Dimerization process, 19
Dioxin, 112–13, 138–39, 152
Divalent mercury, 163
DNA photolyase, 14f, 19–23, 28–29, 73
Dose-response relationships, 4

Ecological risk assessment, 173n2
Ecotoxicology, 157
Electron transfer, 50
Elements, essential, 4, 13–14, 53–57
Endocannabinoids, 102
Endocrine disrupters, 8, 107, 110, 166–68
Endocrine system, 5–6, 101, 167–68
Endogenous chemicals, 102
Endogenous ligands, 199n43
Environmental imprinting, 125
Environmental response genes, 116f, 120
Environmental signals and endocrine systems, 167–68
Environmental stressors, 29, 46–47, 58–59, 116f, 122
Environmental stress response (ESR), 120–21, 126
Epel, David, 124–25
Epigenetics, 125
Essential metals, 56
Essential minerals, 5, 53–57
Estrogen, structure, 111
Estrogen receptors (ER), 102, 104–5, 110, 143
Eukaryotes, 24, 53, 66f, 68–70, 188n1, 189n10
Evolution: adaptive, 133; convergent, 141–42; of CYP detoxification enzymes, and plants, 82f; of detoxification systems, 1–2; of DNA photolyase, ultraviolet light and, 14f; gene family, 93; sulfur chemistry and, 52–53. See also Contemporary/rapid evolution
Evolutionary adaptation, 5, 30
Evolutionary arms race between prey and predators, 94–95

Evolutionary mismatches, 10–11, 25, 46–47, 81, 149
Evolutionary perspective, 1–3
Evolutionary rate estimates, 147
Evolutionary theory, 23, 135
Evolutionary toxicology, 133, 143
Extinctions, major, 144, 150–51

Federal chemical regulators, 152–53
Fenton reaction, 39–40
Ferric iron (Fe^{3+}), 38, 50
Ferrous iron (Fe^{2+}), 37–38, 50
Feyereisen, Rene, 93, 96
Fish, 139, 142–43, 204n20
Flame-retardants, 154
Flavodoxins, 55
Fossil fuel combustion, 163, 169
Frogs, 27–29
Fruit flies, 60, 77–78, 113
Furanocoumarins and CYPs, 95–99

Gasch, Audrey, 120–21
Gatekeeper genes, 72, 75, 79–80
Gene alleles, retention of, 64
Gene conservation, 105
Gene duplication, 22–23, 59–60, 105, 110, 146
Gene-environment interactions, 125
Genes: general responders, 120–21; lateral transfer, 137; number in organisms, 20; p63 and p73, 76, 78–79; regulation of, 19; renegade, 72; response, 122; up-regulation and down-regulation of, 142; XPA, 24. See also DNA photolyase; P53 gene family
Genetic drift, 23
Genetic mutations, 15–16, 73–74
Genomics, defined, 20
Giant hogweed, 95
Glucocorticoids, 108
Glutamate receptors, 88
Glutathione, 43
Goldstone, Jared, 94–97, 110–11, 117, 122–23
Granites, 51
Gray, L. Earl, 107–8, 110

Great Oxidation Event (GOE), 37, 51
Green algae, 85
Green chemistry, 158

Hamdoun, Amro, 124–25
Hartung, Thomas, 9, 152, 156
Hemoglobin, 41–42
Hendry, Andrew, 133, 136–37, 143–44
Heritable resistance, 140–41
Hillman, Bob, 72–73
Homeostasis, 5–6, 56, 117, 119–20
Hormesis, 8, 45, 126–27
Humans: capacity to overwhelm evolutionary adaptation, 30; and changes in metal availability, 64, 164; and changes in stratospheric ozone, 25; gene complement, 119; impacts of, 149–50; and perturbations that induce evolutionary changes, 10; role in microevolution of life, 147; UV-induced damage in, 24
Hydrogen peroxide (H2O2), 38–40, 42
Hydrogen sulfide (H2S), 52
Hydrothermal vents, 55
Hydroxycoumarins, 96
Hydroxyl radical ((OH), 39

Immune stress, 59
Induction, 112
Industrial chemicals, 107–8, 111, 152–53
In vitro research, 121–22
Iron, 42, 49–50, 52, 54–55

Kelce, Bill, 107–8
Killifish, 138–39, 141–42, 145
Kinnison, Michael, 133, 136–37, 140–41, 143–44
Kirschvink, Joseph, 36, 38
Klerks, Paul, 62–63
Krewski, Daniel, 155–56

Ladino clover, 89
Lane, Nick, 33, 37, 69–71

Last universal common ancestor (LUCA), 21–22, 41, 60, 113, 183n30
Latex, 87
Lead poisoning, 6
Legacy chemicals, 152, 161
Levinton, Jeffrey, 62–63
Life on Earth: altered relationship between UV light and, 25; contaminants available to, 3–4; early warning system, 126; location of evolution of, 20–21; oxygen and, 34; sulfur chemistry and evolution of, 52–53
Ligands, 102–4, 112, 168
Lipophilic chemicals, 106

Macroevolution, 135
Magnetic field, 17
Magnetotactic multicellular prokaryotes (MMP), 189n10
Mammalian NER proteins, 24
Mammals, placental, 22
Marsupials, 22
Martin, William, 70–71
McLaughlin, John, 167–68
Mercaptans, 163
Mercury, 127, 162–65
Metal-binding proteins, 55–57, 60
Metallochaperones, 191–92n38
Metallome, 55–57
Metalloproteins, 42, 80
Metallothionein (MT): conservation of functional structure across phyla, 64; in defense against cadmium, 58, 62; genealogical lineage of, 59; in the human body vs. mice, 61; role of, 57–60, 80–81; trade-offs, 81
Metals: changes in availability and new selection pressures, 50; changes through life's evolution, 48f; and contemporary evolution, 62–64; essential-to-toxic continuum, 53, 56; estimating availability of, 53–54; global redistribution and environmental challenges, 50–51;

history of, on Earth, 49–50; homeostasis in, 61; human-initiated changes in availability, 64, 164; interaction with oxygen, 35; oxygen and availability of, 51–52; replacement of one for another, 56; varying susceptibility in different species, 60
Metazoans, 66f, 71, 76, 105, 117, 123–24
Methicillin, 134
Methyl mercury, 163
Mineralocorticoids, 108
Minerals, essential, 5, 51–57
Mithridates, 83, 90
Mitochondria, 42–43, 68–69, 172, 183–84n38, 188n1
Molecular exploitation, 110
Molecular genetics, 7
Molecular oxygen, 34–36
Mollusks, 109, 113
Molybdenum, 54
Monarch butterflies, 90
Montreal Protocol on Substances That Deplete the Ozone Layer (1987), 26, 165
Mosquitoes, 145
Multicellular life, 69–70, 93, 121
Mutations: cancer-promoting, in p53 gene family, 79–80, 191–92n38; in DNA, 15–16, 19, 73–74; natural selection and, 23; somatic, 76–78
Myxobacteria, 189n10

Nacci, Diane, 141–42, 144–45
Nanochemistry, 147
Nanoparticles, 161, 168–72, 204n29
National Research Council (NRC) Committee on Toxicity Testing and Assessment of Environmental Agents, 155
Natural selection, 5, 10–11, 23, 136, 138–44, 146
Nematodes, 77–78
Nuclear receptors (NR), 100f, 104–7, 108

Nuclear transcription factors, 113
Nucleotide excision repair (NER), 24, 74
Nucleotides, 16
Nutrition, toxicology and, 4–5

Omics, 9, 121–22, 155
Organophosphate (OP) resistance, 145
Orphan receptors, 106
Ouzounis, Christos, 21–22
Oxidation, 35, 49–50, 52
Oxygen: atmospheric, 4, 36–37, 92; atomic, 180n3; basics, 34–35; controversy over early history of, 36; evolution of, 36–39; and the evolution of catalase, 32f; initial production of, 17; and life on earth, 34; microorganisms which flourish in absence of, 40–41; molecular, 34–36; toxicity of, 39–40; UVR and, 38. See also Reactive oxygen species (ROS)
Oxygen detoxification, 33–35
Oxygen-evolving complex (OEC), 37
Oxygen stress, 43–47
Ozone, ground level, 150–51
Ozone, stratospheric, 16–19, 25–26, 165–66

P2X, 103–4
P2Y, 104
P53 gene family, 3, 66f, 68, 71, 76–81
P63 gene, 76, 78–79
P73 gene, 76, 78–79
P450 enzymes. See Cytochrome P450 (CYP)
Pathways of toxicity (PoT), 155, 156
PCB congeners, 161–62
PCB resistance, 141–43
PCBs, 112, 138–41, 145, 152, 161–62
Penicillin, 134
Peppered moths, 134, 202n3
Perfluorinated chemicals, 152, 154
Peroxidases, 42–43
Pesticide resistance, 97, 145–46

Pesticides, and oxidative stress, 43–44
Pharmaceuticals, modern, 84, 88
Phenotypes, novel, 125
Phenotypic plasticity, 140–41
Photosynthesis, 17–19, 37, 181n10
Phylum *Loricifera*, 189n4
Plants, 82f, 84, 85–91, 95
Plasticity, 140–41
PM10, 44, 46
Polyaromatic hydrocarbons (PAHs), 73, 112–13
Polybrominated chemicals, 152
Polybrominated diphenyl ethers (PBDE), 154
Population genetics, 70
Population studies, 157
Prokaryotes, 24, 68–69, 189n10
Protein encoding, 122
Protein folding, 56–57, 172
Protein receptors for ATP, 103
Protostomes, 86
Puffer fish, 90
Punctuated equilibrium, 135
Purines, 103
Pyrimidines, 19–20, 103

Rapid evolution. *See* Contemporary/rapid evolution
Reaction norm, 140
Reactive oxygen species (ROS): abiotic-derived, defenses against, 39–40; air pollutants and generation in lungs, 44, 46; detoxification of, 41; DNA integrity and, 73; flexibility of response by antioxidant systems, 45; generation by nanoparticles, 171; H2O2 and, 38; as hazardous byproduct of aerobic respiration, 42; mitochondrial production, 183–84n38; oxidative stress and, 47; singlet oxygen, 35
Receptors, 102–4, 110–14. *See also specific types of receptors*
Redox electron transfer reactions, 50
Reproductive receptors, 107–8

Resistance to toxins, 63, 139, 141–43, 145–47, 204n20
RNA nucleotides, 15–16, 53

Saxitoxin, 87–88
Science (journal), 2–3, 151–52
Secondary plant metabolites, 88, 95
Selective pressure, 50, 61, 63, 84, 121, 137, 146, 152
Selin, Noelle, 164–65
Siderophore production, 55
Silent Spring (Carson), 157
Smolen, Michael, 133, 143
Somatic cells, 73–74, 76–78
Speciation, 135
Sponge genome, 71, 77, 189n11
Standing genetic variation, 63–64, 138, 147
Starfish regeneration, 78
Starlet sea anemone, 77
Stegeman, John, 96–97
Steroid hormone receptors, 108–11
Steroids, 89
Sterols, 92, 98
Stratospheric ozone, 16–19, 25–26, 165–66
Stress response, 120–21, 126
Stromatolites, 181n10
Structure-activity relationships (SARs), 112
Sulfate-reducing bacteria, 164
Sulfur chemistry and evolution of life, 52–53
Summers, Kyle, 67, 75
Sunlight, 16–20, 25–27, 36. *See also* Ultraviolet *entries*
Superoxide dismutase (SOD), 41–43
Superoxide-dismutase enzymes, 56
Synthetic chemicals, 151–52, 161–62

Thiols, 163
Thornton, Joseph, 101, 109
Toxic chemicals, 4–6, 111–12, 134, 157–59
Toxic defense, 3, 5
Toxicity: first documented cases, 6–7; mechanisms in nanoparticles,

171; of mercury, 163; pathways of, 155, 156; predicting, 112, 153; underlying cause of, 6
Toxicity testing, 7, 155–56
Toxicogenomics, 156
Toxicology: challenges of, 154–55; defined, 1; of drug and chemical metabolism, 2; evolutionary, 133, 143; evolutionary history of, 6–9; goal for twenty-first century, 157; and nanoparticles, 170; nutrition and, 4–5; occupational, 7; paradigm shifting in, 9, 155, 172; traditional, 7–8, 118–19
Transcriptome, 142

Ultraviolet A (UVA), 16
Ultraviolet B (UVB), 16, 23–24, 26–28, 39, 74, 165–66
Ultraviolet C (UVC), 16–17
Ultraviolet light, 4, 14f
Ultraviolet radiation (UVR), 15–16, 16–19, 25, 38, 52, 177n16
Umbelliferae plant family, 95

Verkhratsky, Alexej, 101, 103
Vertebrates, 78, 86–87, 105, 114, 136, 145
Vinclozolin, 107–8, 125, 197n20

Water, and atmospheric oxygen, 36–37
Water-soluble chemicals, 5
World Meteorological Organization, 166
Worms, 60, 62–63, 109

Xenobiotic-activated receptors (XARs), 115
Xenobiotic-binding receptors, 111–12
Xenobiotics, 114
Xeroderma pigmentosa (XP), 24, 74
XPA gene, 24

Yeast, 119–20

Zinc, 53, 55, 80–81, 191–92n38
Zinc-binding proteins, 55–56

5/12 ∝